A craft approach to technology

R. W. Barnes

County Inspector in Craft Education, Essex
Previously Assistant Inspector of Schools, Birmingham

G. C. Knight

Teacher of Design, Crafts Centre, Birmingham

Macmillan

First published 1974

Published by
MACMILLAN EDUCATION LIMITED
London and Basingstoke

Associated companies and representatives throughout the world

Printed and bound in Great Britain at The Pitman Press, Bath

Acknowledgements

The authors wish to thank their respective wives for the help they have given in producing this book

Robert Barnes
Grahame Knight

Contents

	Mechanisms	page 1
	Structures	14
1	Measuring the wind	28
2	Measuring weight	35
3	Measuring temperature	43
4	Measuring air pressure	48
5	Pumps	53
6	Wind power	63
7	The theodolite	72
8	The periscope	78
9	The rangefinder	82
	Materials	85
	Appendix	89

Foreword

It is not only for practical reasons that children at school should learn something of technology. For them its impact will be inescapable. It will be part of life, especially for those whose work or hobbies directly involve technology. What then is the challenge to schools?

Educationally, technology has the merit of purposefully bringing together many disciplines and skills to give practical applications to theoretical concepts; an exercise that would be demanding for the most able children. They will not *learn* technology but, if so inclined, they can develop an attitude towards it on which a special and deeper study can be based. It also has the function of encouraging an *understanding* of technology, not only in practical terms but, also in its social implications, its economic and aesthetic consequences and health hazards. No course in technology is complete without giving consideration to the overall effects of technology.

It is not claimed that creative craftsmanship is in itself a technological subject. It is a contributory element to the technicalities of some but not all of the innumerable technologies, and in some contexts can suggest what technology is about. Practical application of theory encourages realistic attitudes; and although it is often overlooked, a need for investigation into principles and theories can be aroused through intelligent study.

Technology is a word loosely used and often misunderstood. At an intellectual level it is the application of fundamental knowledge to practical things; it should have a degree of originality and be a creative experience. It is not a technical process or a skill: these are parts, but it is adaptation, modification, innovation and concept that challenge the teacher of craft. They provide opportunity for his skills to contribute to better making, efficiency, economy and production, all of which are vital to technological advance. This is the craftsman's way to technology and the basis of consultation across subject disciplines. Educationally it extends the craftsman's range, from making the set pieces with considerable freedom to making with the discipline of productive method, to limits of cost and time and the control exerted by resources and materials. This gives craftwork another dimension and the technological approach becomes part of Craft Education.

It is necessary for one to see the association between craft and technology to know what each can give to the other. We know of distinguished technologists who began as craftsmen and learned their technology through the experience of making. Many others, less distinguished, working at the level of technician, have *thought* as technologists and from their knowledge of working processes and the function have conceived significant improvements through skill and understanding. There are those also with a more academic background who find it necessary to know processes and materials to complete an integration of theory with practice. Yet craft is not technology and technology is not making technical things, and the challenge to the craft teacher is to evolve a method of teaching that is both creative and restrictive—freedom for ideas but limited by time and complexity. This does not mean only the making of things mechanical and scientific but that the terms of making are prescribed not always as a restrictive drawing but to a written brief with 'omissions' for involvement, giving an opportunity to think about the *way of making* and not only about the *processes of making.*

This development does not conflict with traditional craft methods. It is equally exciting and enriches the content of Craft Education. Good craft teaching should stimulate by presenting problems to solve. They need not be technological; indeed, where they are contrived to allow involvement with other disciplines it would be a mistake if total emphasis on the technological and technical aspects of craft obscured its cultural value for those attracted to things beautiful and well made. But technology need not do this. It has its own cultural background in biography, history, industrial archaeology and the remarkable achievements of things made by man whatever the limitations of tools and materials.

Technology has been described as 'scientific usefulness' and in that the craftsmans' role is indisputable. The danger lies in becoming too ambitious too soon and confusing complexity, prestige and technical expertise with creative opportunity. This book shows how from the beginning, through modest projects, the right attitude of thinking can be encouraged and so give technology a place in Craft Education.

Edgar Heelas
Staff Inspector (Schools)
Birmingham

TO THE TEACHER

This is an ideas book, written for pupils. It is organised on a topic basis and includes activities, investigations and necessary background information leading from the consideration of materials and mechanical principles to the making of working models, such as a weighing machine, a water pump, a thermometer, suitable for use in actual situations. Work within each topic section varies in difficulty and the sections themselves progress from simple to advanced, thus the book is useful for pupils with a wide range of interests and abilities. Sufficient assistance is provided to allow the pupil to work at his own pace without actually copying a particular design. Suggestions for further developments of the topics are given in the Teachers' Book. The overall aim of this approach is to encourage individuals or teams to solve problems by employing their own inventiveness, resourcefulness and planning.

The manner in which the book is presented ensures that it is sufficiently flexible to be of value in a variety of situations, for example: Mode Three examinations, and with respect to providing stimulating work for school leavers;

as part of normal examination work; and as a branch of craftwork itself. The activities call for discussion, written work, drawing (both freehand and in accurate detail), investigation and testing, and practical work with familiar tools and materials.

There is variation in presentation from section to section because of the different treatments required, but the headings give broad guidelines for the pupil to follow.

TO THE PUPIL - *HOW TO USE THIS BOOK*

1. Read the sections on Mechanisms and Structures. Carry out some of the activities to test that you understand the basic principles.
2. Select a topic section, read it and carry out the investigations and tests. Choose which type of machine you are going to make.
3. List the conditions which the machine you have chosen must fulfil (for example, must it be waterproof, light in weight, heat resistant? What are the safety requirements?) The list will then form the brief from which you can work.
4. Consider how much time is available, what materials can be used; have you sufficient knowledge and skill for the job? Then investigate and experiment, make notes and sketches, and make prototype models.
5. Produce an accurate drawing of your machine.
6. Construct the machine.

It is useful to keep sketches, notes and mock-ups.

Where you require help consult other people, books, magazines, drawings, and study objects with similar mechanisms.

You are advised against examining in detail any single previously-made example of your type of machine; instead, consider a wide variety and investigate the merits of each one, developing your work from that stage. Look for faults and shortcomings in commercially-produced articles and try to avoid them.

Mechanisms

Section 1

LEVER AND FULCRUM

The lever is one of the oldest and most commonly used devices for gaining a mechanical advantage - that is, moving a load with less effort than would be necessary if the load was moved directly without any other help. The fulcrum is the pivot about which the lever moves.

Unfortunately, this gain is offset by the need to move the effort further than would be necessary if the load was moved directly.

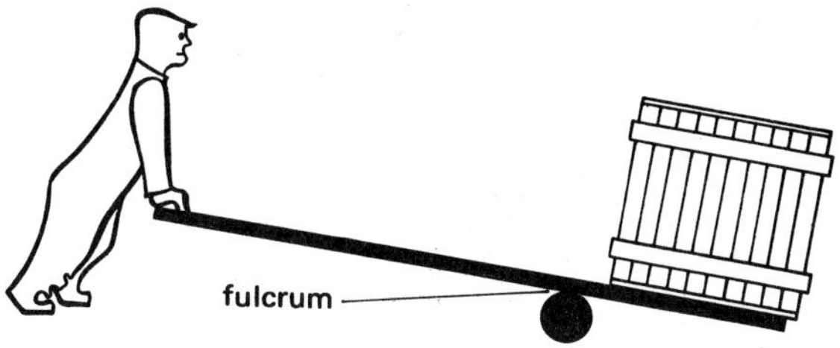

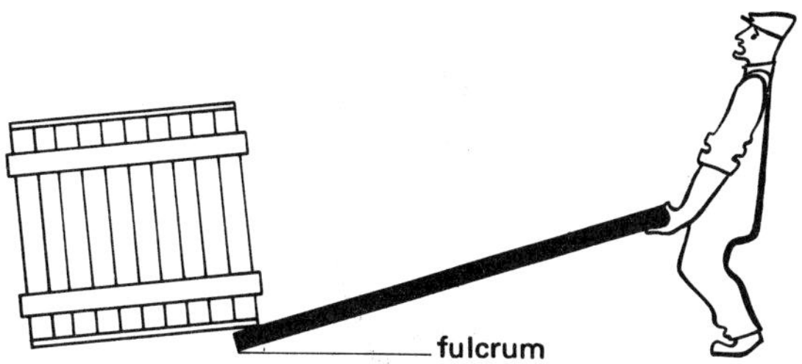

Tongs and nut crackers both use the principle of the lever. The grip exerted at the handles is magnified at the working end.

When a movement is magnified (as on a pointer), the effort applied has to be greater than the load. In an aneroid barometer, for example, the aneriod capsule changes slightly in length - but with a relatively large force.

This minute movement is magnified by a thread over a pulley in order to move a pointer through a distance which can be easily observed.

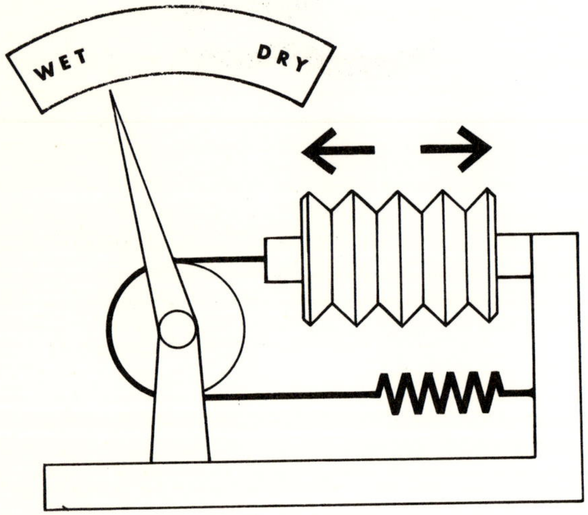

THE SCREW THREAD

A screw thread is a round wedge. By moving the screw thread through a great distance, but with only a small effort, a powerful force can be applied at the end of the thread. In the clamp shown the force exerted by the screw on to the wood in the clamp is greater than the force applied to the tommy bar. However, the effort has to move quite far in order to compress the wood (the load) through a small distance.

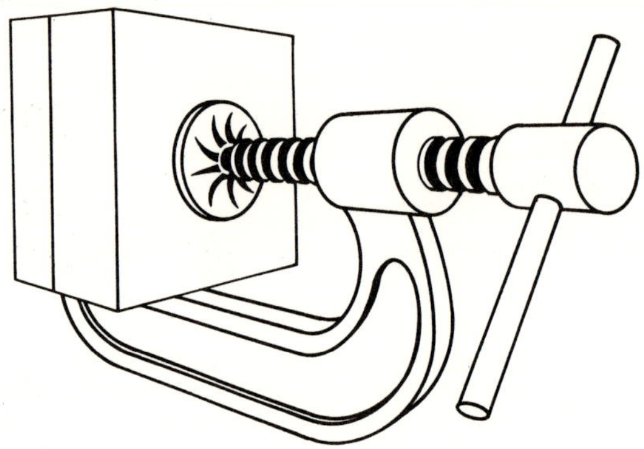

PARALLEL LINKS

Frames, made from links of equal length pivoted at the ends, can vary in shape but their opposite links remain parallel.

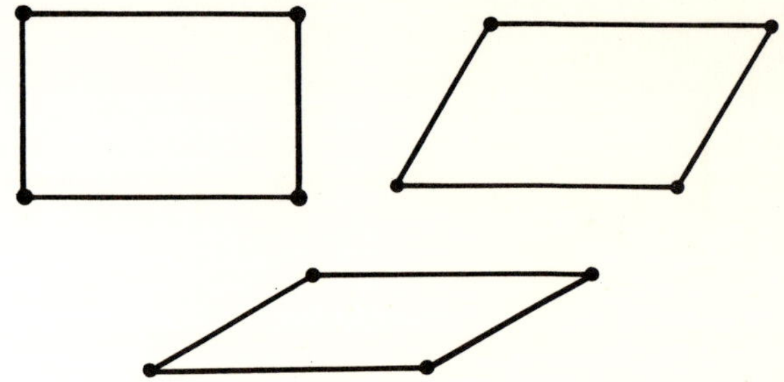

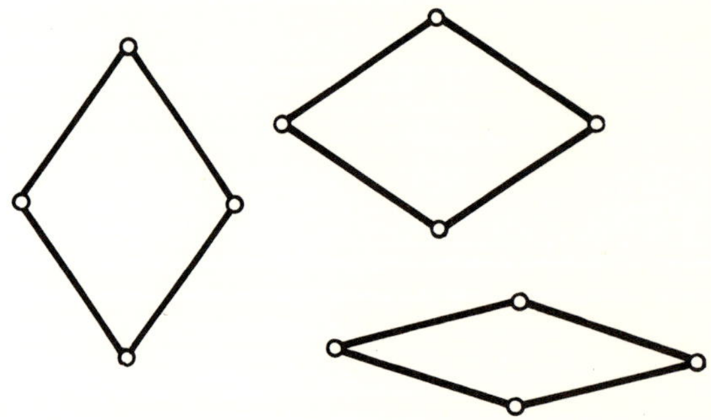

This applies for any position as long as there are four links. Here are some applications. Think of some others.

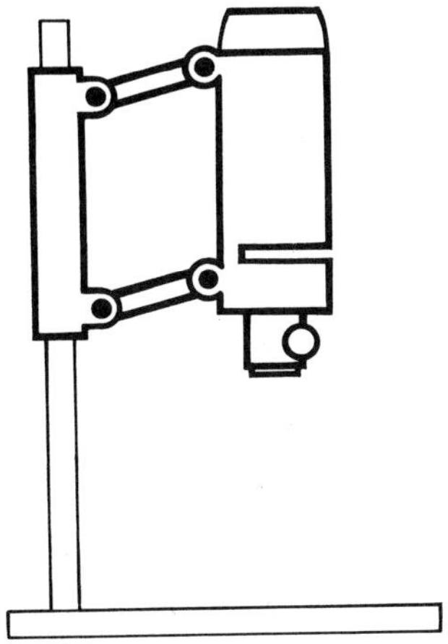

Enlarger

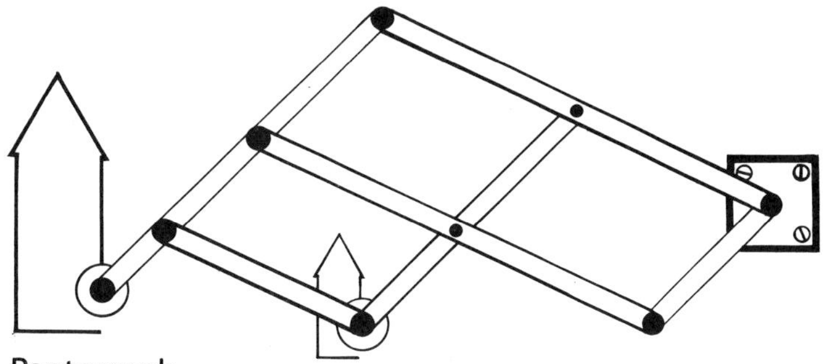

Pantograph

Navigator's parallels

ACTIVITIES

1. Sketch the lever system of a bench shearing machine. By measuring the positions of the pivots and the length of the arm find its mechanical advantage.
2. Sketch the screw thread of a vice. Find its velocity ratio by working out how far the end of the tommy bar has to travel in order to move the jaws 25 mm.
3. Sketch the handbrake system of a car, or the braking system of a bicycle. Does the mechanism on a bicycle increase the mechanical advantage of the hand pulling on the brake lever?

Design and make

a simple nut cracker suitable for a junior-school child to use with one hand.

Two related problems are involved here.

(a) To find the force needed to crack a tough nut.

(b) To find the force exerted by a small hand.

Your solution will incorporate the findings of both investigations. Simple test rigs will be needed to find the answers to the problems.

Section 2

TRANSMITTING POWER AND MOVEMENT

There are many occasions when it is necessary to transfer motion from one part of a machine to another. A car, vacuum cleaner, washing machine and lathe are a few examples of machines that need to do this to operate.

THE PULLEY

A pulley can be used to change the direction of action of a rope, cable or chain.

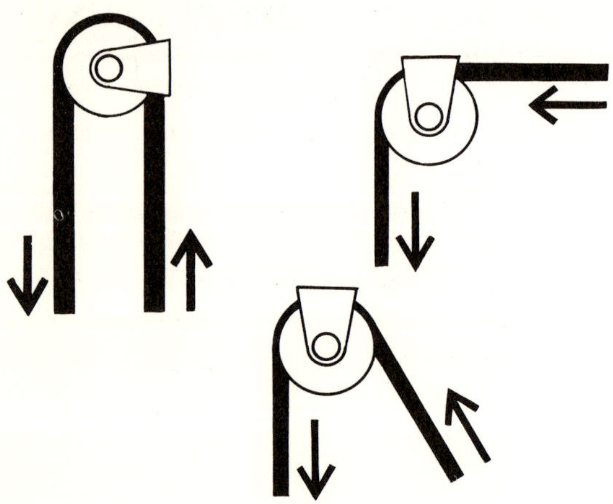

It can also be used to gain mechanical advantage.

If there were no friction it could lift a load by applying an effort equal to half the load, a mechanical advantage of 2 (see the drawing below left). The effort has to move twice the distance moved by the load.

The system below right has a mechanical advantage of 4. The effort has to move four times the distance of the load.

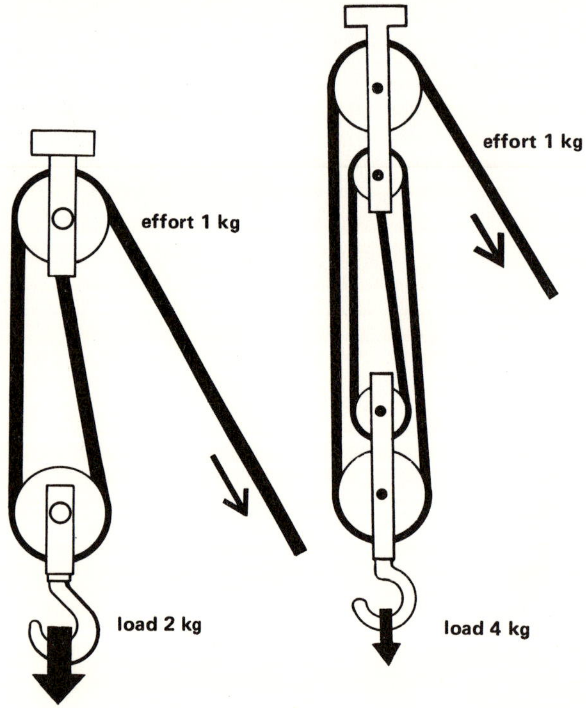

POWER TRANSMISSION

Pulleys can be used to transmit power from one shaft to another, in forward or reverse, or vary the speed between shafts, giving increased mechanical advantage or increased velocity ratio.

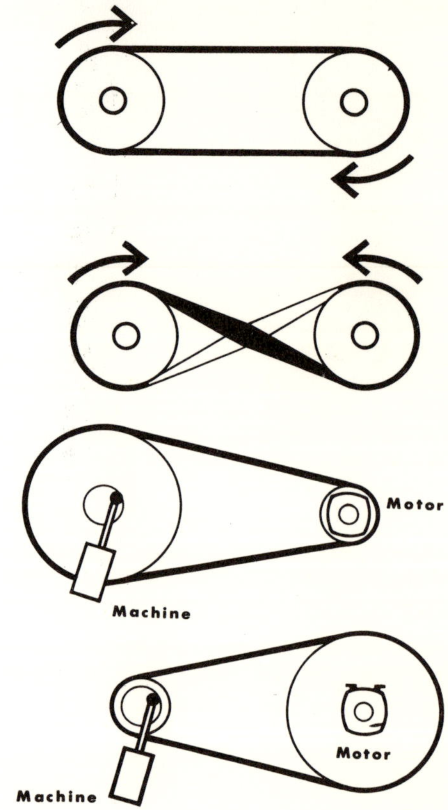

There are two main types of belt and pulley: the flat belt and the vee belt.

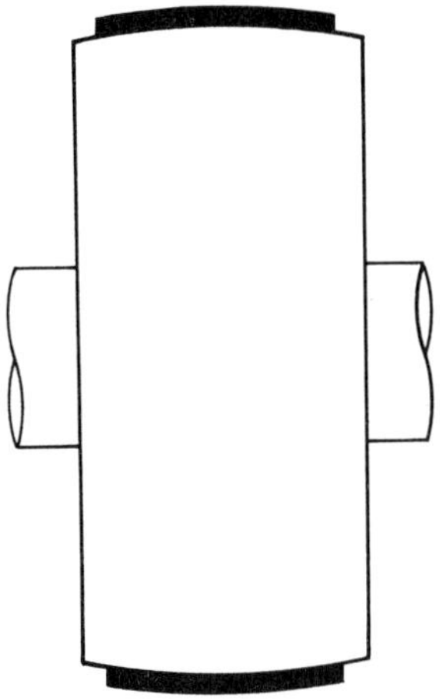

Flat belt

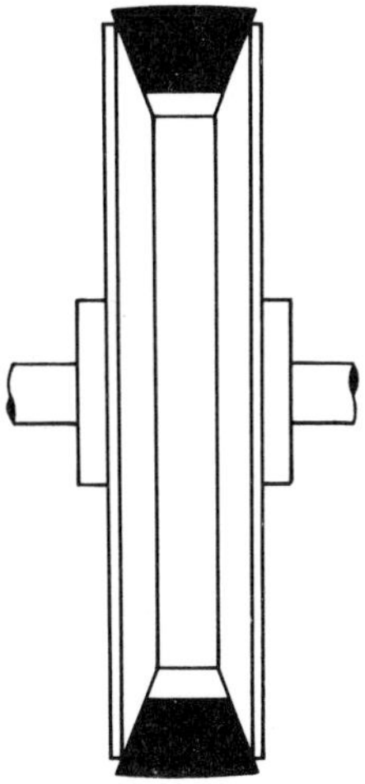

Vee belt

There are variations on these, often designed to provide a positive or non-slip drive or an easy method of lengthening or shortening the belt.

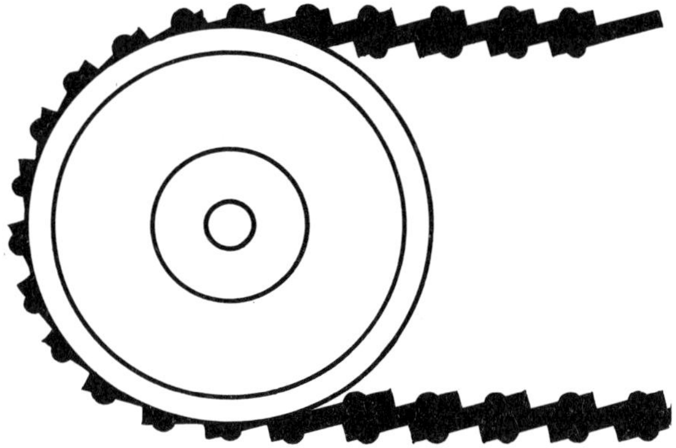

Link belt

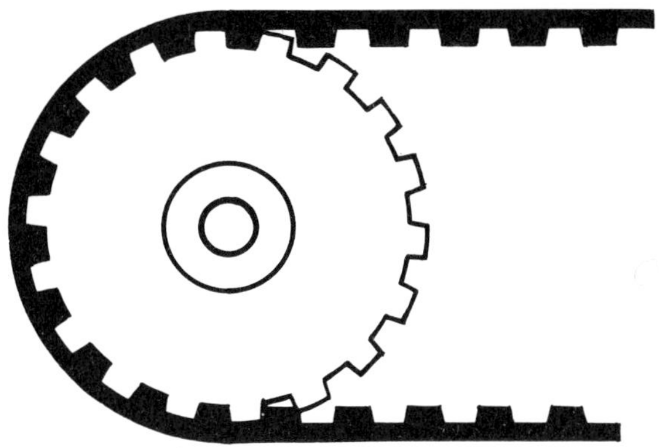

Toothed belt

GEARS

In addition to the mechanical advantage, gears can change the direction of rotation, the speed ratio (velocity ratio), the direction of drive, or all three. Identify the functions performed by each of the gears illustrated here.

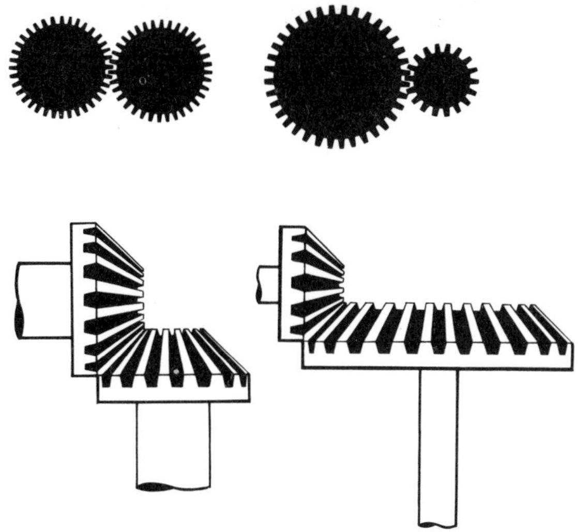

RACK AND PINION

This is a gearing system in which one gear takes the form of a straight line.

Its uses include:

the driving system on a mountain railway where ordinary wheels would slip;

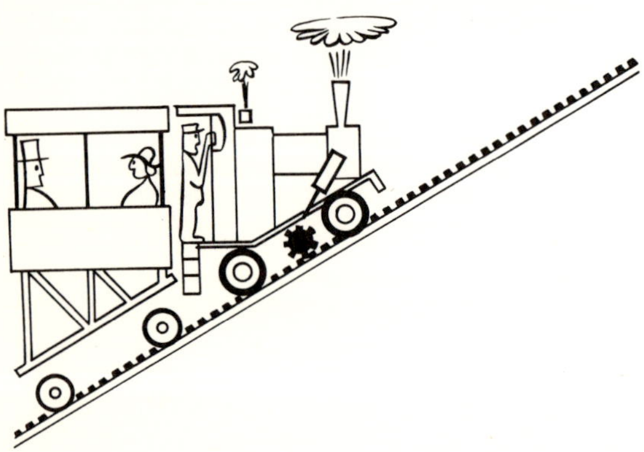

the means of moving the carriage along a lathe bed;

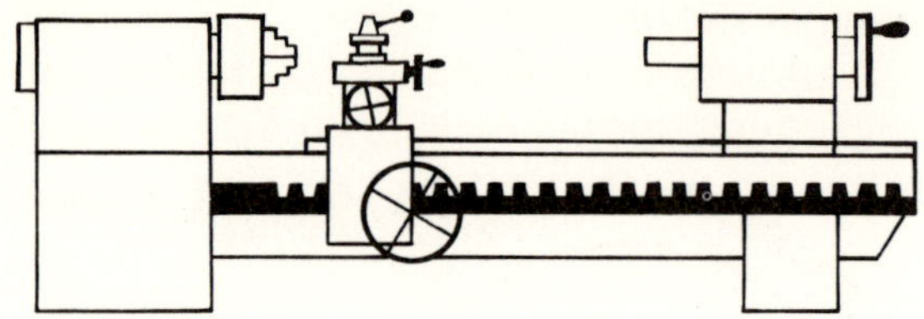

the steering system on some cars in which it is designed to turn rotary motion into a straight line movement.

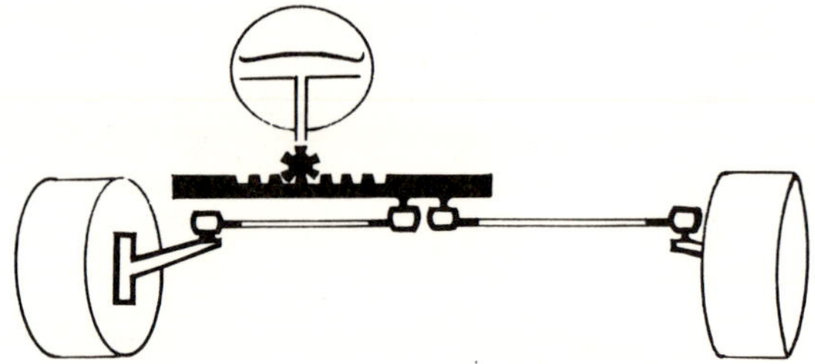

Note The lathe rack would normally be placed with the rack teeth downwards – why?

CAM AND FOLLOWER

A cam and follower offers another method of changing rotary motion into straight line motion. The motion can be of varying speeds or of differing mechanical advantages.

Cams are also used for producing a constantly repeated movement (a cycle).

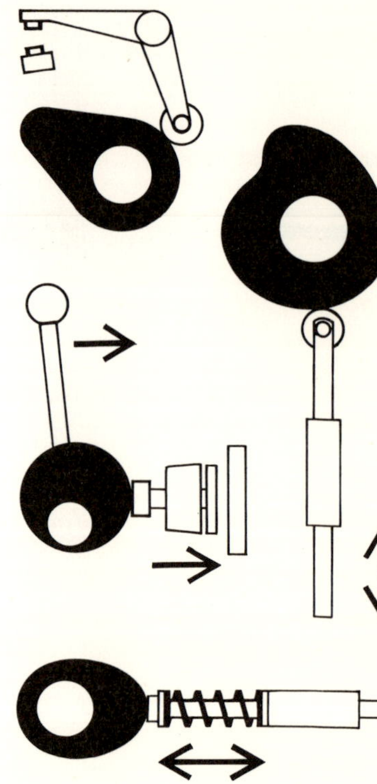

ACTIVITIES

1. Sketch the pulley system on a power drilling machine or on a lathe. Show how the speeds can be changed.
2. Give one example of a rack and pinion mechanism to be found on a device used in schools. Describe the function it performs.
3. Sketch one example of a gear train, showing enough detail for its mechanical advantage and direction of rotation to be seen clearly.
4. Describe one device which uses a cam to move parts — such as the bobbin winder on a sewing machine; or the contact breaker on a motor cycle; or a motor mower engine valve mechanism.

Design and make

Using small (meccano type) gears, design and make one of the following:
(a) a device for winding-in: a fishing line; a clothes line; or control lines for model aircraft;
(b) a simple gearbox for a model steam engine which would allow a choice of forward or reverse directions at crankshaft speed;
(c) a paper punch using a cam system to move the punch.

SLOT MECHANISMS

An inclined slot can be used to produce a required movement.

Taper turning can be done accurately on the lathe using an attachment incorporating this type of mechanism.

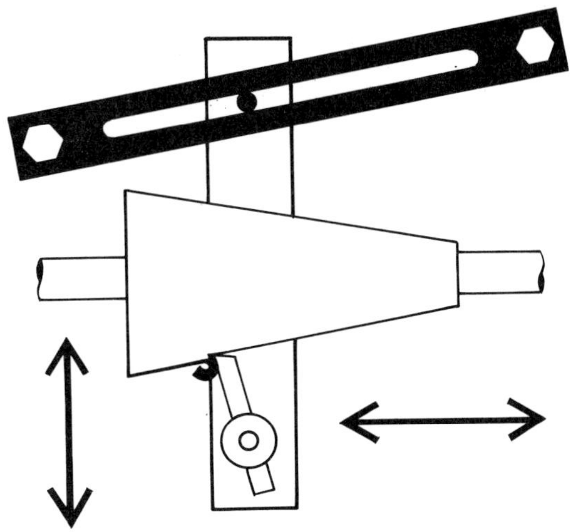

Some lock mechanisms make use of the inclined slot to move two or more bolts. The lock in the photograph could be made in several different ways.

THE WHITWORTH QUICK RETURN MECHANISM

This is a slot (or slideway) mechanism which is used in some machine tools. You might find it used on the shaping machine in your metalwork room. There is a distinct difference in speed between the cutting (forward) stroke and the return stroke.

Which direction of stroke supplies the greatest force to the tool?

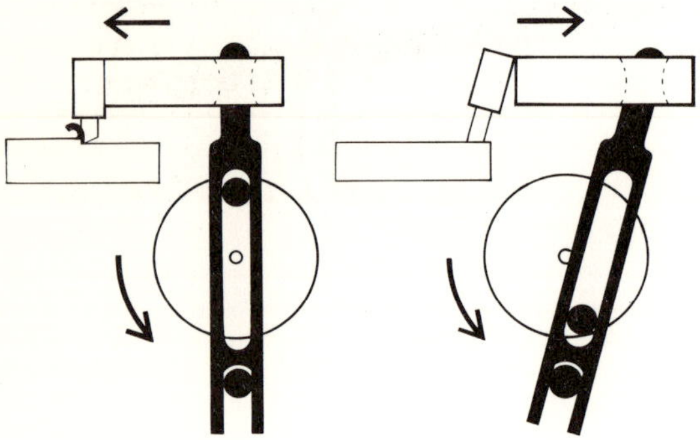

HOOKE'S JOINT

This is a device which allows two shafts to be joined. It enables the turning motion of one shaft to be transferred to the other even though they might not be in alignment under all conditions.

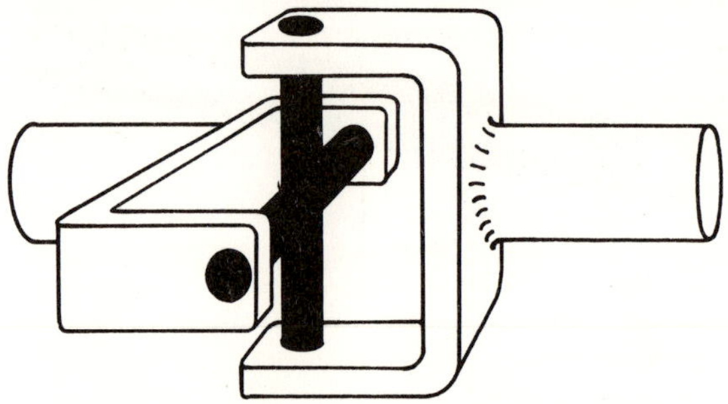

Two of these 'universal' joints are used between the gear box and back axle of a vehicle.

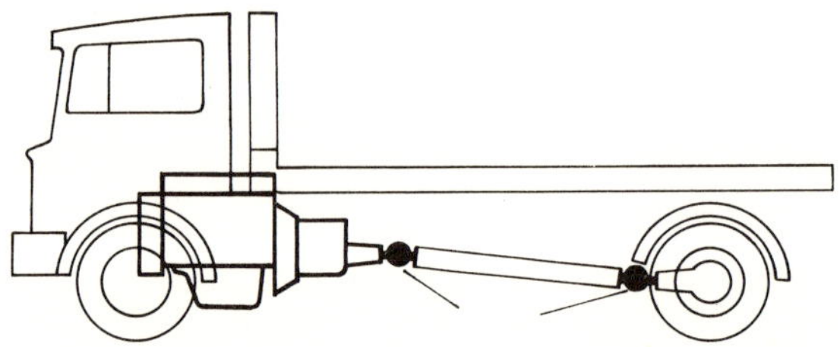

RATCHET AND PAWL

This is a device used to allow a wheel to turn in one direction only or to push a wheel round one tooth or more at a time. Sometimes the ratchet has more than one pawl.

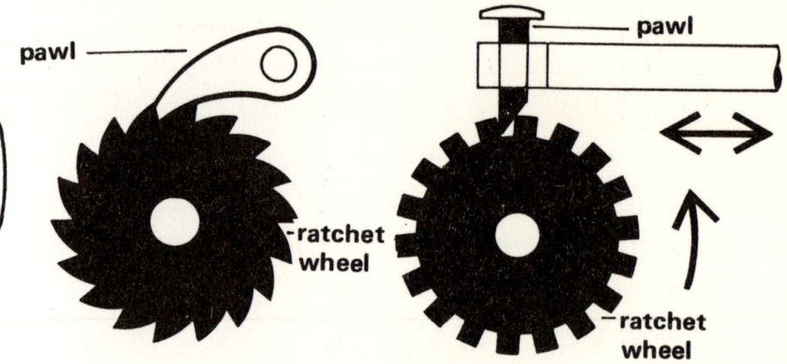

The drawing shows a bicycle free wheel mechanism. In fact it is not a **true** free wheel but a ratchet device (we will look at an actual free wheel mechanism below). Three pawls engage ratchet teeth as the sprocket is turned clockwise by the bicycle chain. When the sprocket is stationary the pawls can slide over the ratchet teeth allowing the inner section to rotate clockwise so that the rear wheel of the bicycle can 'free wheel'.

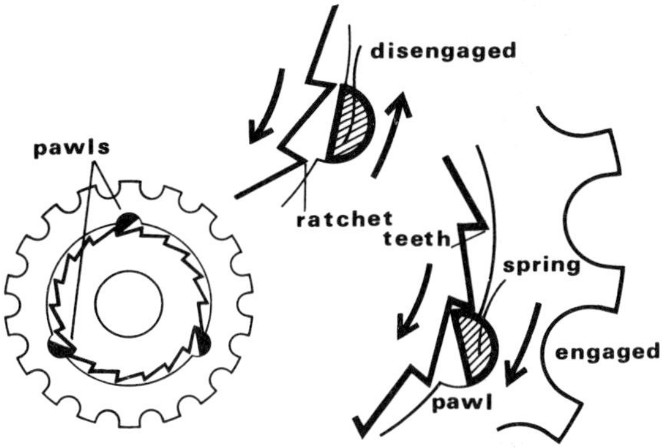

FREE WHEEL

The diagrams show a true free wheel mechanism. When the outer component is moving clockwise, the three ball bearings jam themselves between the inner and outer components. This makes the inner one move with the outer. When the outer component moves anticlockwise, the ball bearings are freed and the inner component is no longer forced to rotate.

SPRING CLUTCH

This grips when one half of a shaft rotates in a direction which causes the spring to wind up and tighten on the other.
Like a free wheel, the part being driven can rotate faster than the driver if forced to do so by a secondary driving force acting in the same direction as the first force.

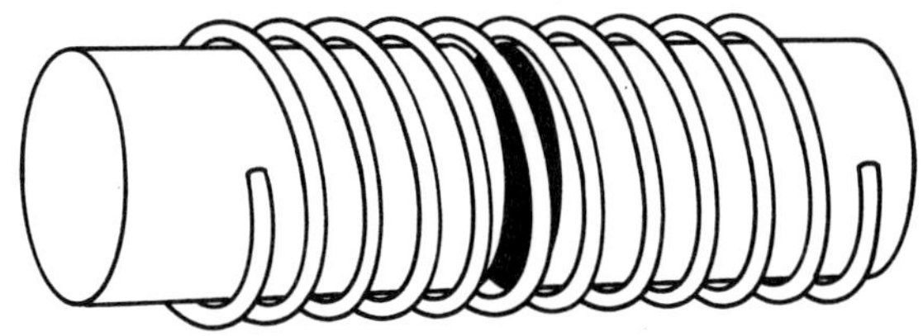

THE WINDLASS

This is a device, sometimes power-driven, sometimes animal-powered, which helps to pull a very large load. It is a combination of a wheel, axle and lever.

A few turns of the rope round the windlass are usually needed to prevent it slipping.

ACTIVITIES

Select some of the following and devise practical solutions.

1. A garage door-stay which will: hold the door firmly open; stop the door opening too far; release easily when the door is to be closed; lie neatly along the door when it is closed.

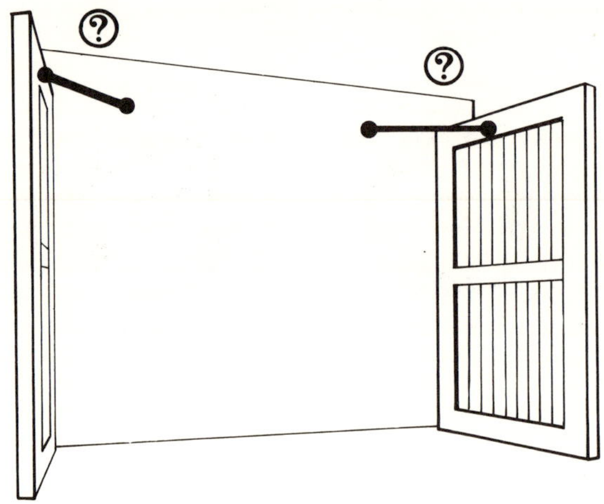

2. A coupling which will allow one shaft to drive another, when they are either: at an angle of between 180° and 120° or with their centre lines parallel but between 1 and 5 mm out of line.

Problems of this sort are encountered on model power boats at the point of coupling the engine to the propeller shaft.

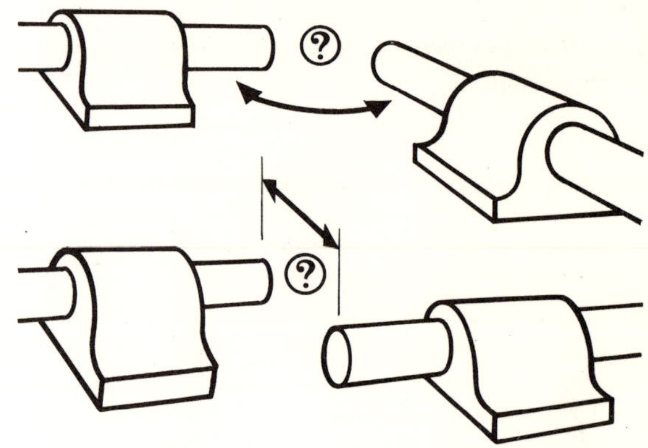

3. A hinge on an animal house which allows a neatly fitting door to open freely in either direction. A plan of the set-up is shown here.

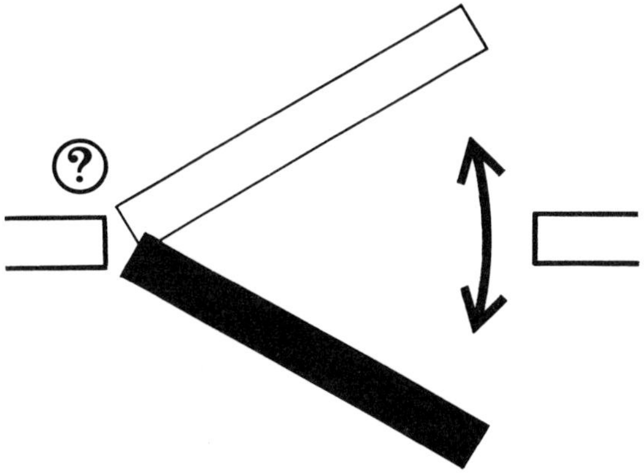

4. A mechanism such as you might fit to a rear window wiper on a car. It needs to be able to convert rotary motion (from an electric motor) into back and forwards motion (for the wiper) at some distance from the motor.

5. A device to move a straight edge freely up and down on a drawing board, keeping it parallel to the top edge.

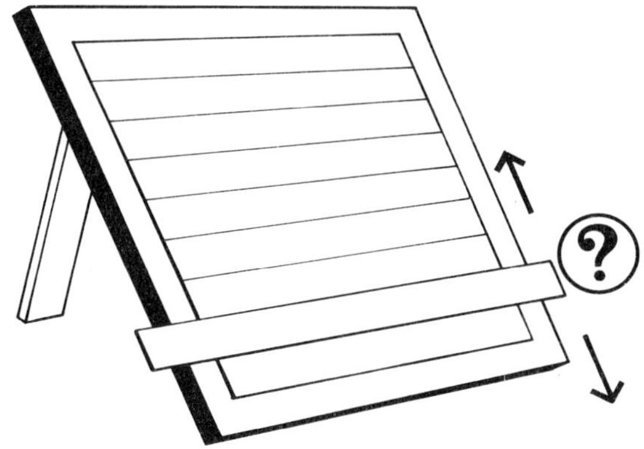

6. A car jack making use of the pull exerted by a screw thread. If you are thinking of using the threaded rod in compression then it will have to be thick. Why?

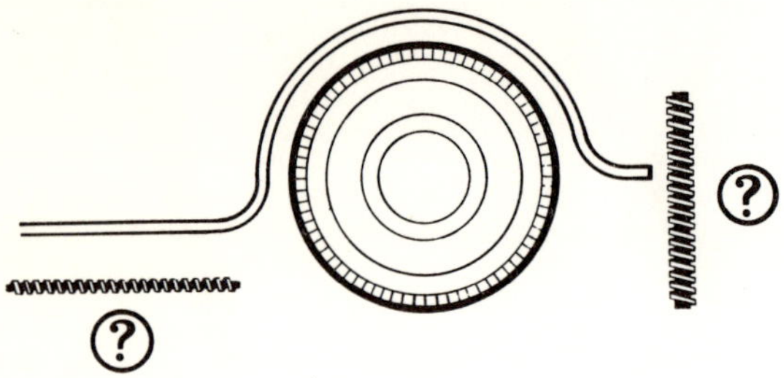

7. A rope and pulley for lifting heavy objects and incorporating a safety device to stop the load falling when the rope is released suddenly.

8. A goose neck for the joint between a mast and boom on a sailing dinghy. It must: allow universal movement; be easily uncoupled; not corrode; slide downward to tighten sail.

9. A lock which will push bolts in several directions round a door or lid so as to make it very secure.

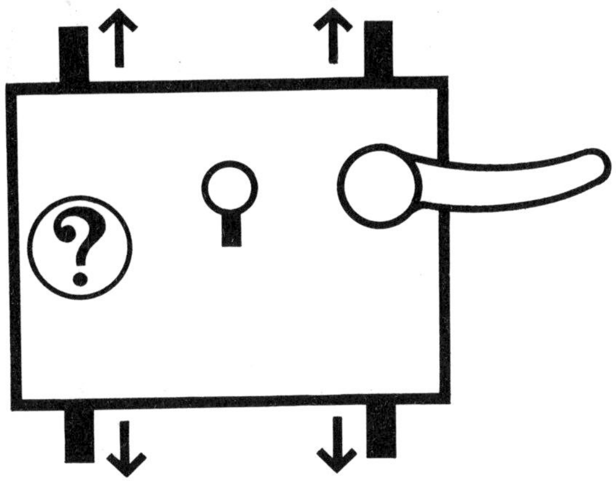

10. A mounting which will make viewing through a telescope very easy. The eye-piece should remain at the same point no matter how the telescope is moved within a field of vision of say 20° in all directions from the central position.

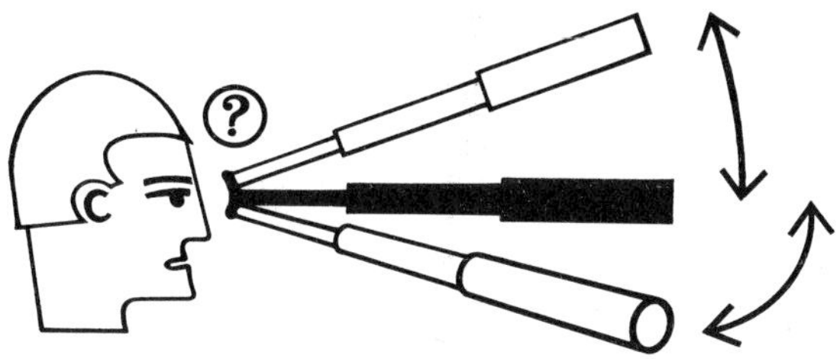

11. A pedal and connecting rod drive for a go-cart having two speed ratios.

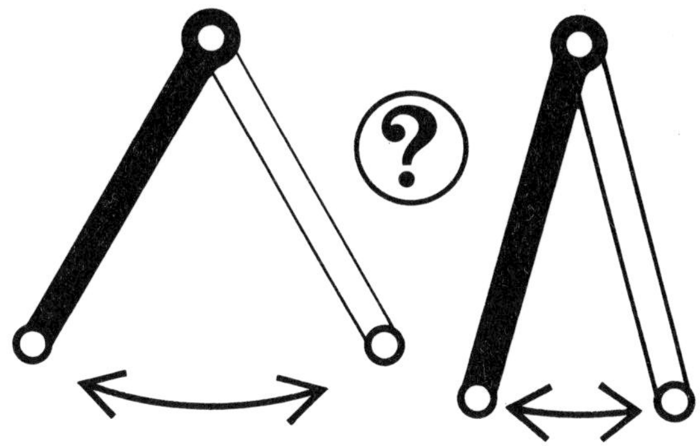

Structures

Section 1

A structure is a device for providing support. Many common objects incorporate a structure; the materials from which it is made and its shape depend on the particular function it is required to perform. Natural structures include the shell of an egg and the bone and muscle of an animal. Much can be learnt from studying many objects, not only by handling them but also by close observation under a microscope. The shape of each component of a structure will give indications of how it is able to carry loads and withstand blows. For example, the way in which bones are hinged and attached to ligaments and muscles will show the type and direction of force each part is shaped to withstand. The shape of wood fibres will show how each cell makes its contribution to the strength of a tree.

In most man-made articles structural strength is important, although in some cases it is not the most important factor.

It is very useful, and often vitally necessary, to think about the manner in which force might act on an item you are about to construct. Success in any craft project depends on the ability of the object to maintain its correct shape under working conditions. In other words the mechanical requirement of any structure is to successfully resist all reasonable forces imposed on it.

The magnitude (size) and direction of these forces often vary between certain limits. Frequently other factors are important in a structure, examples being appearance, weight, cost, constructional time, resources and durability. Some of these may be more important than others. When thinking about different types of forces encountered by everyday structures it is necessary to consider six main types: bending; compression; tension; torsion; shear and centrifugal force. Some structures are subjected to only one of these, others have to resist two, three, four or even all six under their full working conditions. Descriptions of these forces acting on different structures are given on the following pages.

BENDING

The strength of the plank (or a bridge) must be enough to resist the weight of the wheelbarrow without being bent (deflected) to a dangerous degree. This type of structural member is called a beam. Beams for spanning large gaps have to be stronger than those for short spans, even where the loads are the same.

COMPRESSION

A pit prop resists a compression load. A compression member is called a strut. Miners prefer wooden props because they creak when under a considerable strain, thus giving warning of danger. A long one has to be thicker than a short one to resist the same load without breaking or bending greatly; furthermore, the strength of a strut is greatly affected by the type of fixture used to hold the ends in position.

TENSION

In the drawing, the rope must be strong enough to take the pull caused by the weight of the engine. A member for carrying a tensile load is referred to as a tie, and within reasonable limits the length of the tie makes no difference to the load it can take.

TORSION

Torsion is a turning (or twisting) force and is calculated by multiplying the distance (radius of handle) by the force (push on the handle).

Another example is a spanner tightening a nut (investigate the torsion or torque wrench). This principle is used to ensure the correct tightening of nuts on highly stressed parts of cars and aircraft.

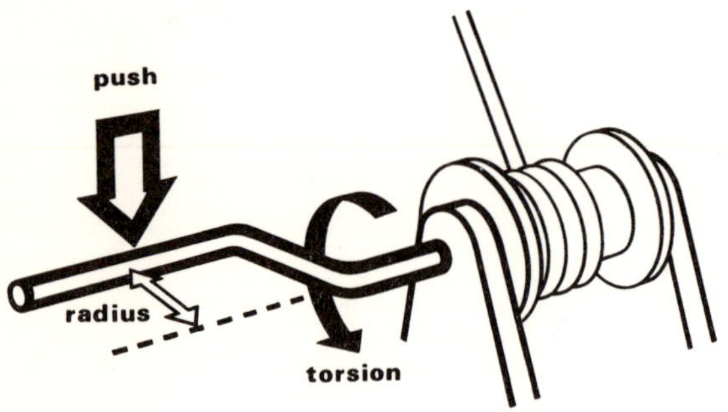

SHEAR

When a material 'shears' it is parted by an excessive force that tends to slide the material apart. In the drawing a trailer hits a bad bump in the road. The effect is of a sudden strong pull on the drawbar by which it is being towed. The force causes the pin retaining the drawbar to shear.

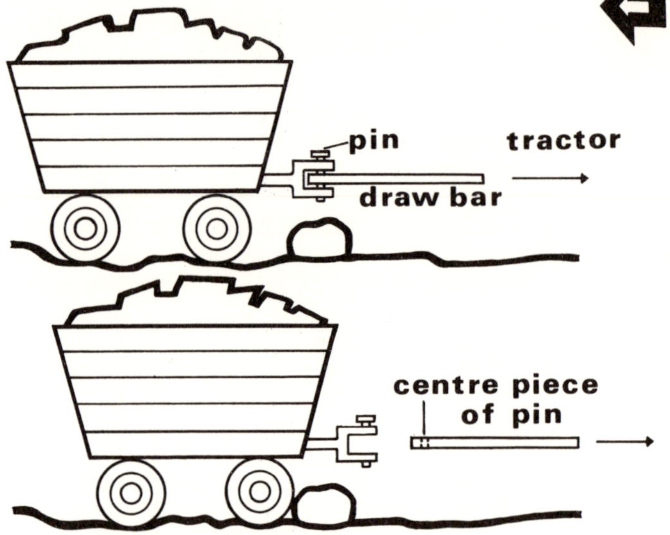

The ruined pin looks like this.

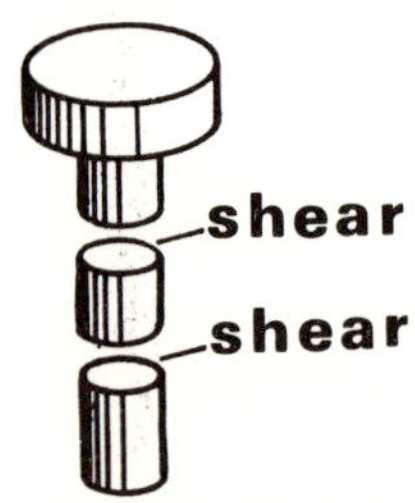

CENTRIFUGAL FORCE

If a weight attached to a length of string is whirled round, the string will become tightly stretched as the weight attempts to break free.

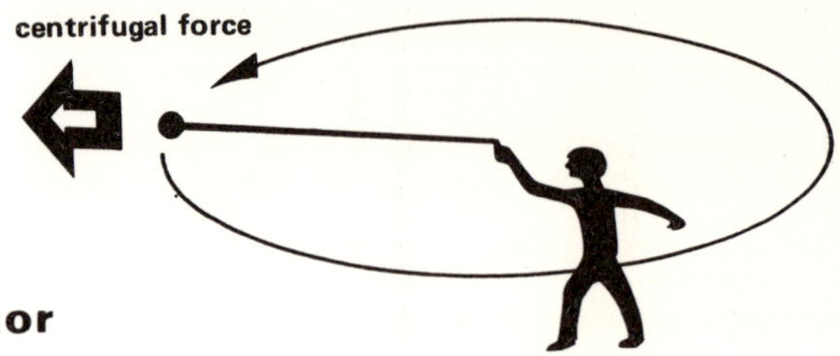

Similarly a wooden disc being turned into a bowl on a lathe will be affected by the rotary motion, each part of the rim will try and wrench itself away from the centre of the bowl. If the speed of revolution is too high the rim will break off, especially if there is a defect in the wood.

The danger of such situations cannot be over-emphasised; this explains why eye shields should always be worn by machine operators. When a particle is chipped off a grindstone it flies out at about 30 metres a second, too fast even for a person with exceptionally quick reactions to move out of the way if he is closer than about 6 metres to the wheel.

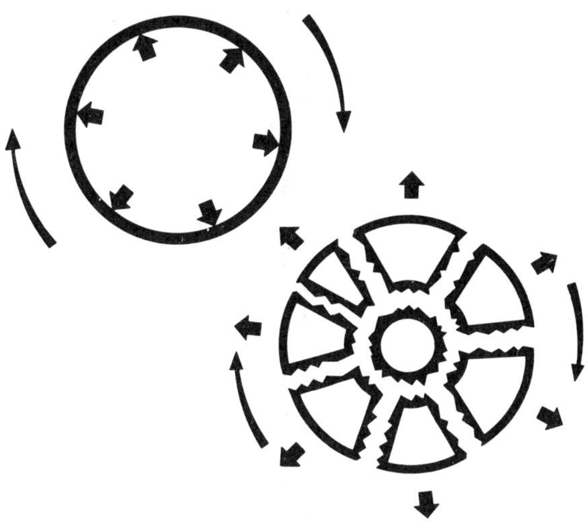

ACTIVITIES

Sketch the examples of structures listed below; state briefly whether the parts are resisting tensile, compressive, shear, bending or torsional loads.

1. A fire extinguisher case with riveted seams.
2. A tap cutting a metal thread.
3. A springboard on the side of a swimming pool.
4. A sash cramp for holding wood joints together while being glued.
5. A car jack.

In all the examples shade in red pencil the part you would expect to break first if the article was overloaded.

Find out

the correct speed for turning (a) a 200 mm diameter wooden bowl, (b) a 50 mm diameter mild steel bar.

Section 2

BEAMS

As mentioned earlier a member used mainly to resist bending is known as a beam.

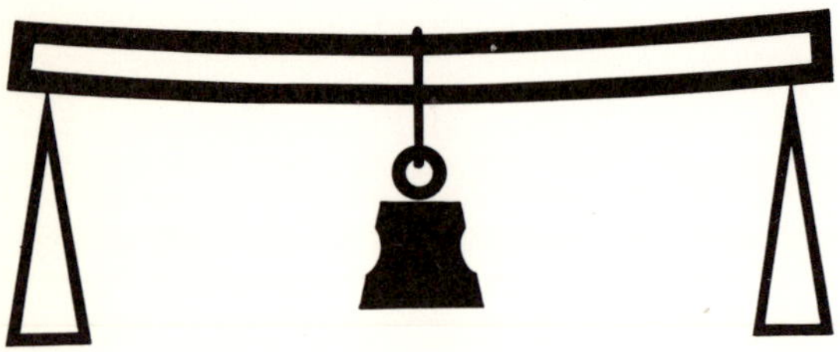

Some simple tests using cardboard beams will show you some of the important facts about beams.

You will find that a deep narrow beam can take a larger downwards load than a wide but thin beam; this is true only if the beam remains in the correct position and does not twist.

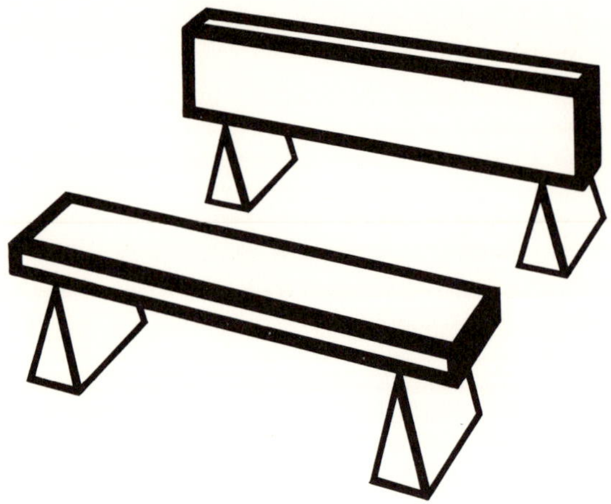

From this it follows that a thin beam will bend more easily than a thick beam of the same length and width if both are subjected to the same load.

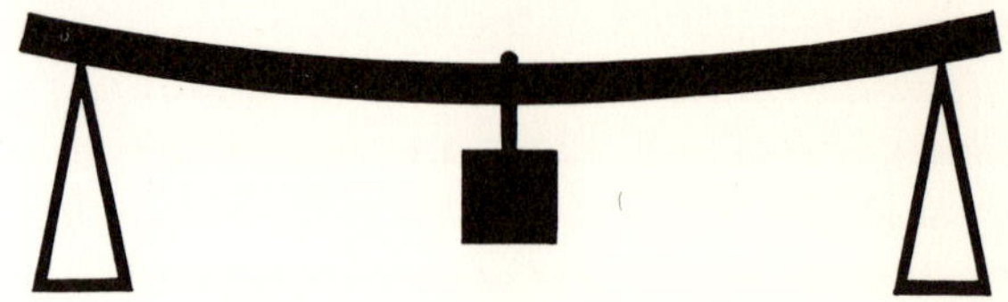

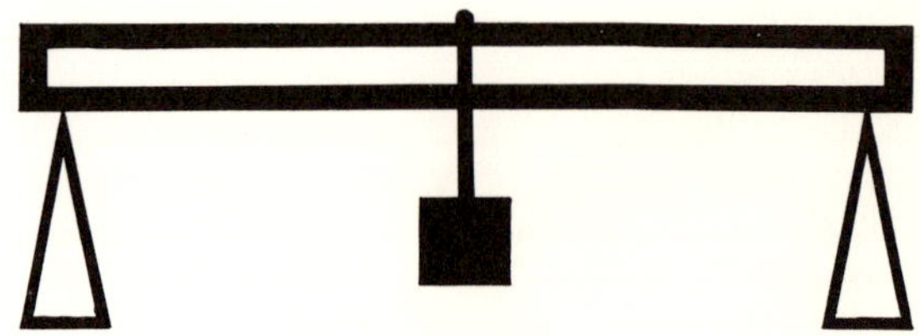

As the span is increased the stiffness of the beam must be increased.

The best way of increasing the strength and therefore the stiffness of a beam is to increase its thickness rather than its width.

Unfortunately any increase in size of a solid beam will result in an increase in weight. This weight increase is a serious drawback. A long beam, for example, will bend considerably under its own weight.

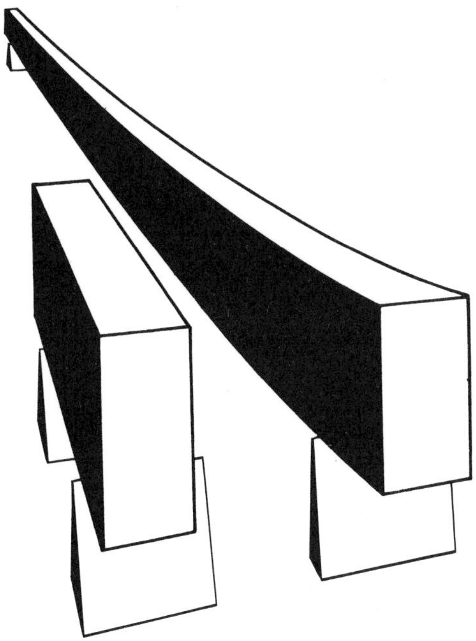

BEAMS IN SPECIAL CIRCUMSTANCES – THE CANTILEVER

A special case is the cantilever beam. An example of this application is a towel rack of the type commonly made in school workshops.

When a towel is hung on the rails the back plate attempts to pivot about its bottom edge. The screw B does very little work, screw A taking most of the tensile load (which tries to pull it out of the wall). The bottom edge takes the compressive force (which tries to press it into the wall).

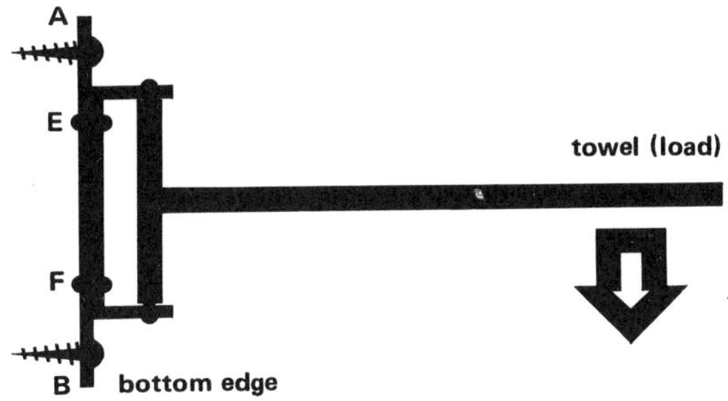

Here is the same situation represented diagrammatically and simplified by altering the direction of A as in lower diagram.

By taking moments, load x Y = force on screw x X. The same principle applies to the rivets E and F. Can you say which of the two rivets is taking the largest load? A simple testing device could be made using this principle to find the maximum tensile loads which different sizes of screws will take when screwed into different timbers.

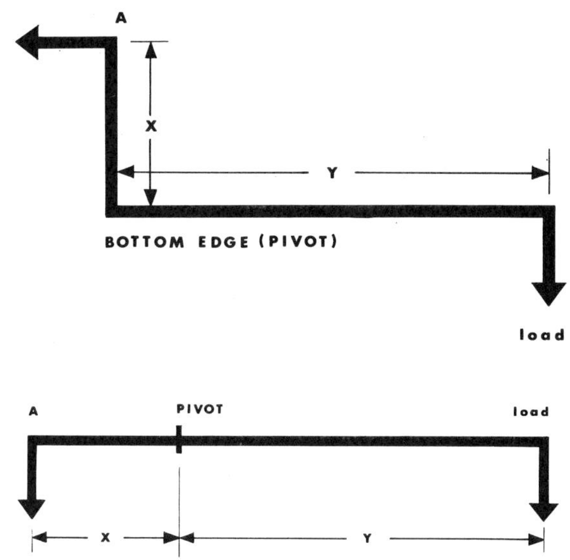

The same situation exists when a door is hinged to a frame.

The top hinge is in tension, the bottom hinge is under compression. The weight of the door also tries to shear the screws.

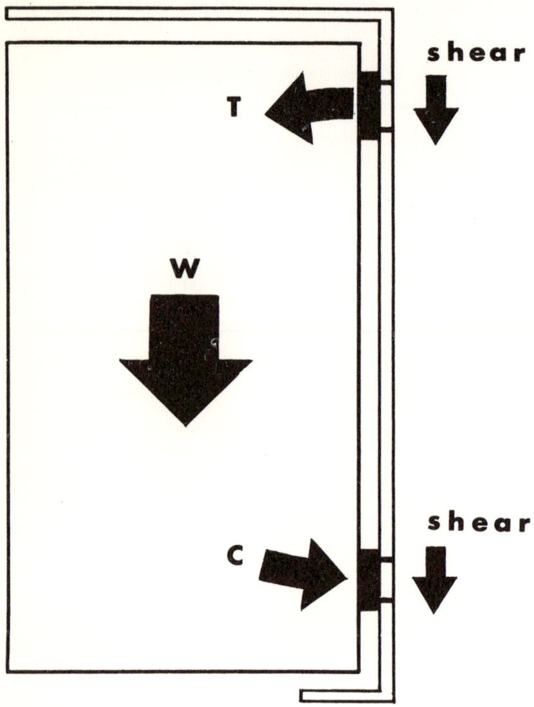

In homes doors sometimes have to be hinged so that they clear carpeting as they are opened or closed. Special hinges are used called 'rising butts'. Because of their shape (see below), rising butts tend to lift the door slightly as it is opened.

Work out why a door will tend to swing closed on its own if fitted with these special hinges.

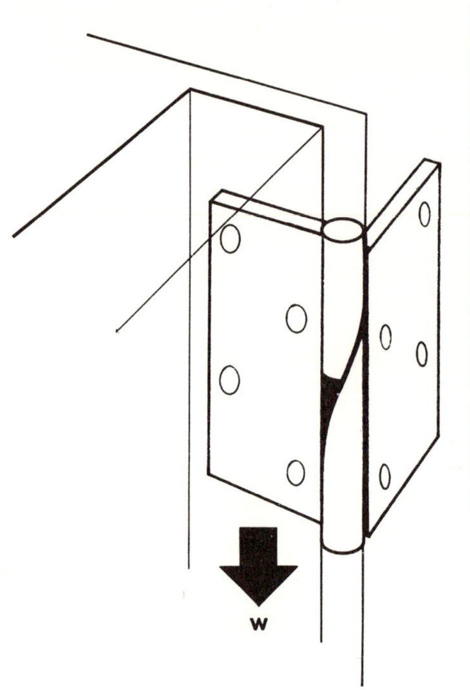

ACTIVITIES

Sketch the following beams and write a short account of why you think the particular shape and material was used. A vague answer such as 'to hold up the roof' will not do. Mention weight, span, fixing, corrosion and rot, expense and maintenance.

1. A concrete lintel over a window or door.
2. A floor joist with a ceiling below.
3. A ceiling joist with no floor above it.
4. A large steel beam spanning the roof of a large hall, railway station or bus depot.

Find out

If the towel rack on page 19 has an arm 1 metre long and the back plate distance AB is 150 mm long what will be the pull on screw A if a towel weighing 1 kilogram is hung on the end of the arm?

Make

1. A device to find out the minimum diameter of a mild steel arm which will carry the towel without bending excessively and which can also be used to find the minimum diameter of a wooden arm suitable for supporting the towel.
2. A device to find the minimum size of wood screws required at A when screwed into a variety of different timbers.

BEAMS IN SPECIAL CIRCUMSTANCES – PILLARS AND STRUTS

Pillars or struts have been described earlier as members under compression. It has also been pointed out that the length of a strut and its fixing have a great bearing on its ability to resist loads. If the strut could be kept straight whilst under load it would be able to take the same load no matter how long it was. Unfortunately, in most cases, due to unavoidable uneven loading, a strut begins to bend as soon as it is loaded. This makes it act like a beam, having its material on the convex side in tension and that on the concave side in compression, the neutral axis of the beam (approximately the centre) taking only a shearing load.

Before making an object some method is required to determine how it might be affected by forces in different situations. This analysis will repay the constructor's effort and time by helping him to anticipate possible causes of failure, and thus to ensure the success of his work.

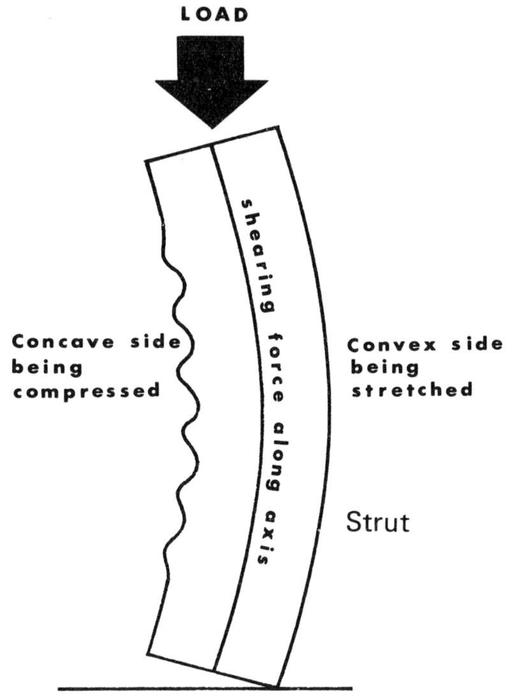

Strut

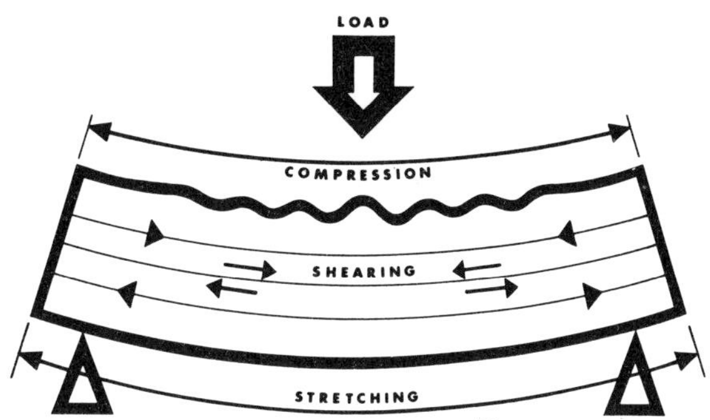

Beam

Section 3

FORCES ON FURNITURE

The forces acting on a chair can change considerably according to the way in which it is used.

Under normal conditions the legs will be under compression, the rails under bending, the webbing of the seat under tension, and the joints between the legs and rails under shear loadings.

If the chair is tilted backwards whilst in use (as chair b is) the distribution of loads will be changed. The back legs take a bending force in addition to a compression, and the front legs are relieved of any force.

When the chair is placed on an uneven floor (see chair c) the stress will be particularly severe on the structural members and joints.

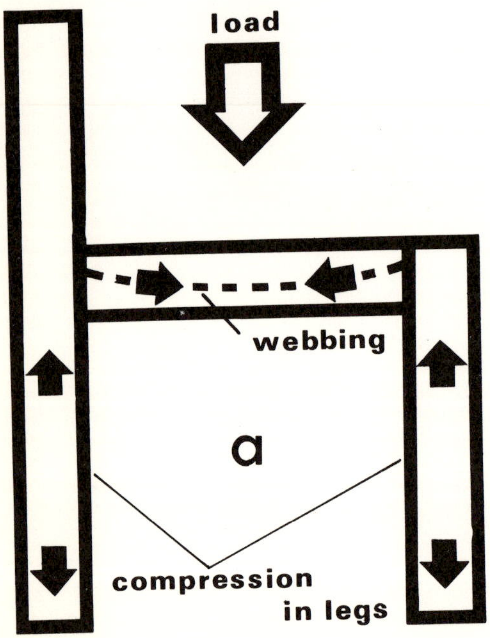

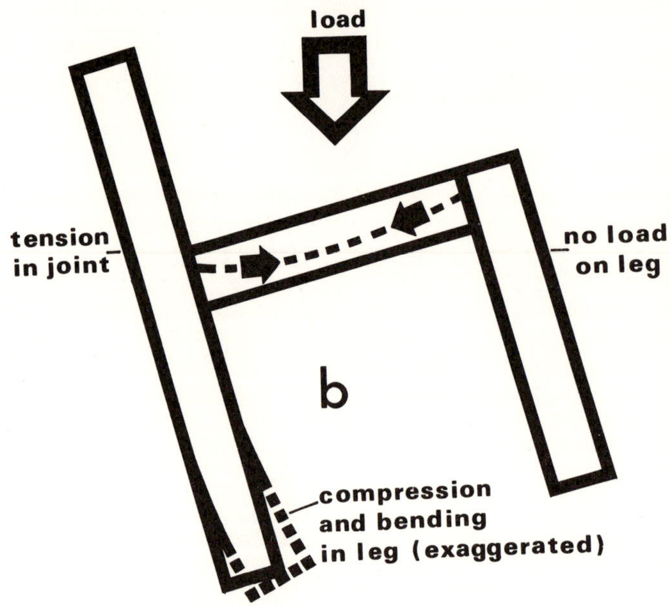

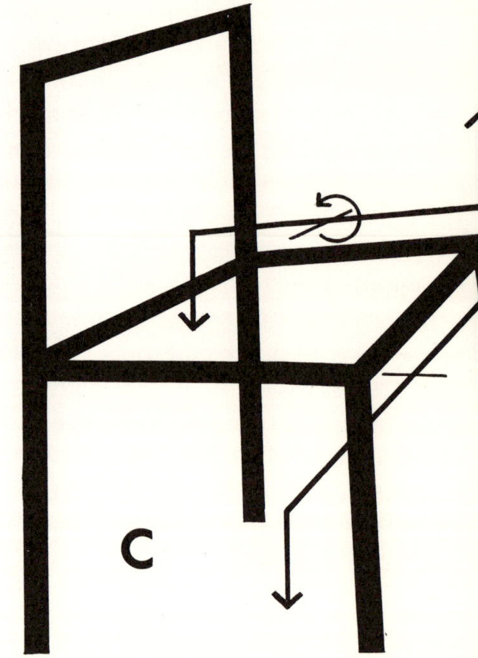

Using the method given previously, here is an analysis showing the main forces acting on the chair.

Position	a	b	c
Webbing	Tension	Tension	Tension
Front legs	Compression	–	Compression
Back legs	Compression	Bending	Compression
Side rails	Bending	Bending	Bending and twisting
Front and back rails	Bending	Bending	Bending and twisting
Front joints	Shear	–	Shear and twisting
Back joints	Shear	Tension Shear Compression	Shear and twisting

A rail, made of a size sufficiently large to prevent it twisting, will cause excessive loads to be thrown on the corner joints. Why is this?

SHEAR

You will have noticed that the shearing forces have been shown as only acting on the joints. This is not strictly accurate as all parts of the rails taking a downward load are in shear. The joints, being cut into the legs and rails, are the weakest part of most structures and for this reason if shearing is to take place it will almost always occur at the joint.

In the chair frame shown the shearing force is trying to separate the tenons from the rails along the line of the shoulder. This force also tries to shear the glue round the tenon and force it out of position. The bottom edge of the tenon resting on the end of the mortise assists the glue in preventing this movement.

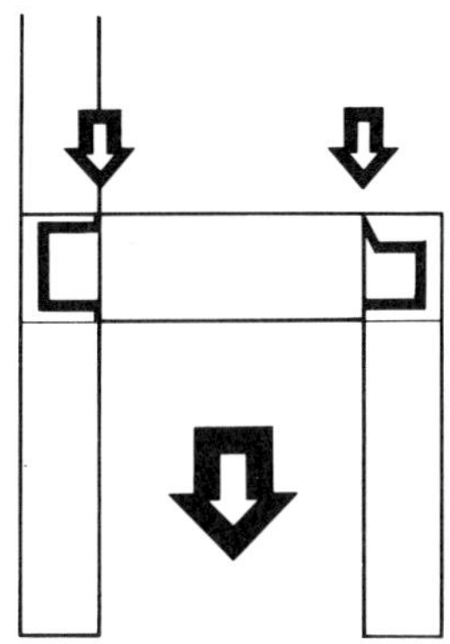

If you refer to the example of the cantilever (towel rack) it will help you to work out the forces acting on the tenon in the mortise in a situation where the chair is tipped backwards.

The arrows indicate tensile force, the longer the arrow the greater the force. In your opinion, at what point is the joint trying to pivot and what is resisting the tensile force trying to pull the tenon out of the mortise?

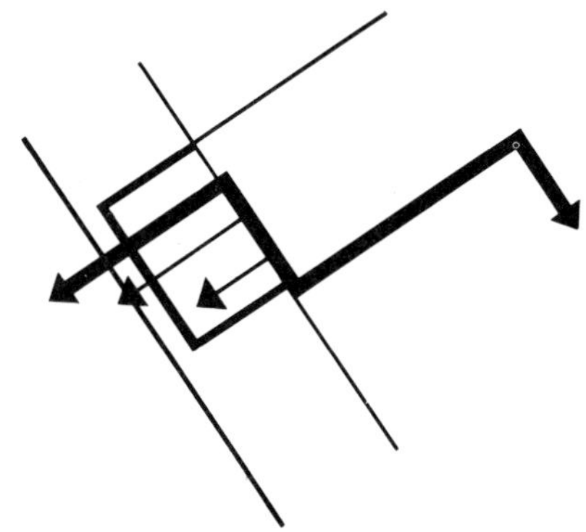

You can see that an accurately fitting joint is best. You should also be able to work out why a deep rail which allows a large tenon to be cut on it is better than a narrow one. Can you suggest what would happen if: (a) the shoulder was badly fitted (b) the sides of the tenon only touch parts of the mortise (c) the joint was starved of glue?

GLUING

In referring to glue and gluing you will have come to some conclusions about the way in which glue works. Some simple tests will enable you to prove that glue is usually much stronger in resisting shear than in resisting a purely tensile force. Some tests are misleading in certain situations. The test shown below is not a good one as it subjects the joint to a progressive failure of the glue through bending. Joints of this sort need screws, bolts or nails to assist the glue.

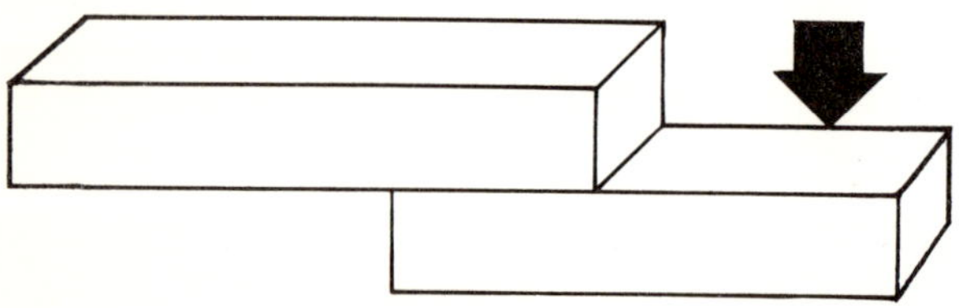

Most glues give excellent resistance to a pure shearing force; a test devised on the lines shown here will be useful.

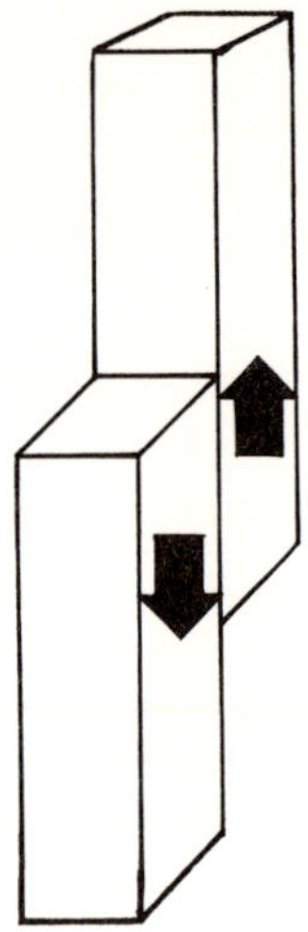

In many situations the glue will be subjected to a shearing force of the type shown where the direction and amount of force is distributed differently over the glued surfaces.
A test of this type, carried out with different glues on identically made joints, will be very useful for making comparisons between glues and the gluing properties of various materials.

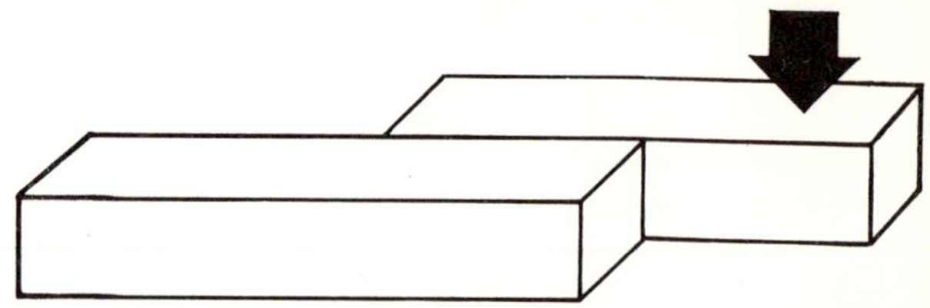

FRAMES

A triangular framework is rigid even if it is only held together with a simple pin or bolt at each corner.

If the members A and B of the framework shown below are deflected under a load they will eventually fail as the load increases.

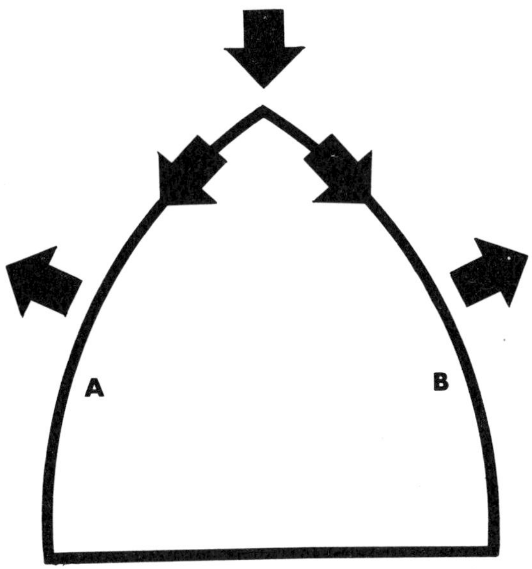

A strengthened corner joint of the type shown below would increase the strength by delaying the tendency of the sides to bend under load.

A square frame would require very strong corner joints or some other method of bracing. What alternative methods can you suggest?

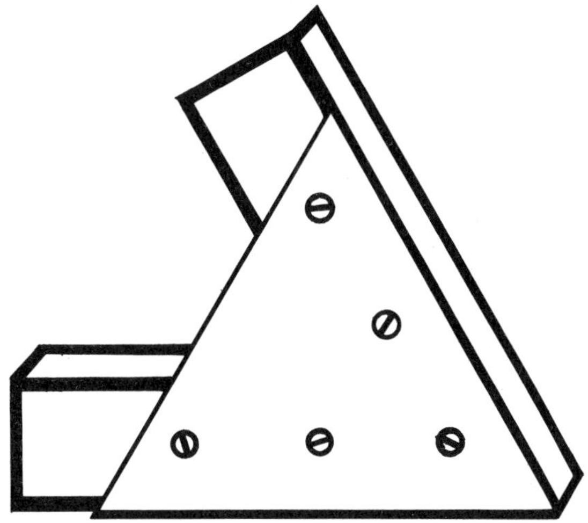

A series of triangular frames can be used to construct a large girder. The great saving of weight is the major advantage of this type of construction.

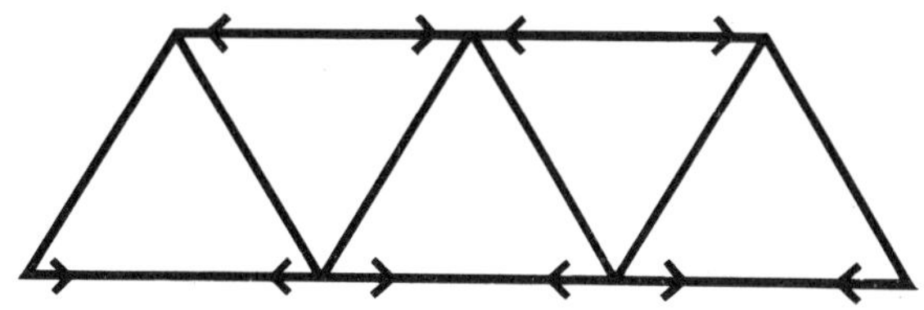

The individual members must be able to withstand the forces imposed on them. A strut is shown by pointing the force arrows away from each other (tension) and a tie with them towards each other (compression).

By experimenting you can find out where the ties and struts are in such a girder. In the same way as in a solid beam, the top surface will be in compression and the bottom in tension.

strut

tie

ACTIVITIES

1. Sketch three types of chair construction used for four-legged dining chairs.
2. Make three different joints for fixing the side-seat rail of a chair to the back leg. Use the constructions from Activity 1. Work with two other pupils. Each of you make a different example. All three pieces should be made from the same size and type of material. Make a simple rig to test the three types of construction — find out how much force they will withstand if under a bending force as shown in chair b.
3. Discuss the results and suggest modifications to improve the two weaker constructions.
4. State the reasons for using stretcher rails in wooden chairs. Indicate the best position for a stretcher under the frame in a four-legged dining room chair.
5. The square frame mentioned on page 25 requires very strong corner joints. Sketch three alternative methods of joining the corners. State your opinion on the advantages of each method.
6. In the girder made from traingular frames on page 25 some of the members could be made from wire or rope. Sketch this beam. Assume it is only to carry a downwards load at the middle and that it has supports at each end and indicate where wire could be used. Say why.
7. Sketch another form of simple girder bear or roof truss. Make a detailed sketch of or of the corner joints.

An example of where you can apply your knowledge and skill in the school workshop.

Suppose you require a rig for lifting engines out of cars. It should be strong, light and able to be dismantled after use. One of many solutions would be a pyramid-shaped structure.

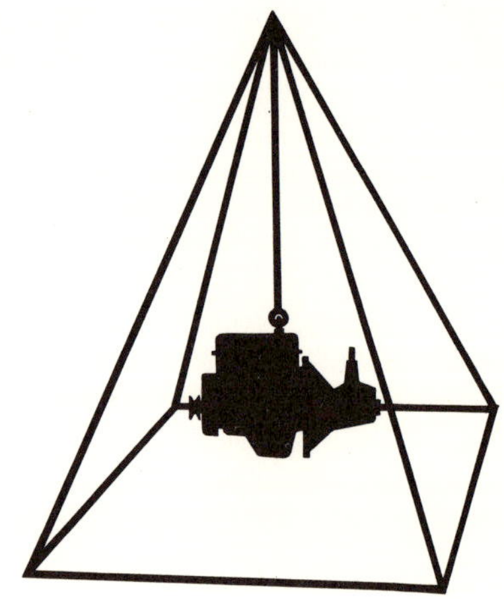

The forces would be:
compression in each leg;
tension in the horizontal ties;
tension in the lifting rope;
twisting of the entire frame.

The legs would also be subjected to bending loads arising from their sloping position and from people leaning on them as they worked on the engine.

Answer these questions.

1. What section, type and size of material can be used for each part?
2. What extra parts could be added to strengthen the frame?
3. How can the suspended engine be prevented from twisting?

A special safety pulley of the type asked for in the chapter on mechanisms would be very useful on this rig. (See page 12 **Activity 7.**)

FRICTION

The ability of one particle to slide over another with ease depends on the friction between the two. Friction is a force which always opposes movement and its magnitude depends on two factors. The first is the coefficient of friction existing between the two materials in contact. In simple terms this is a measurement of the resistance to movement caused by the roughness, hardness and rigidity of the surfaces. Coarse glasspaper against coarse glasspaper has a high coefficient of friction, steel on steel has a fairly low coefficient of friction. The second factor in determining the frictional force is the force which is applied to keep the two surfaces together. The harder you press the glasspaper together the greater the friction. Some situations demand as much friction as possible to be generated, other situations are best served by very little frictional resistance being present. A tyre, brake shoe, tenon and mortise joint all benefit from a high frictional force. A sledge, axle bearing and machine slideway all benefit from having as little friction present as possible. Hardened steel in contact with hardened steel, brass, bronze and nylon give low friction performances especially if lubricated. Steel on aluminium is sometimes used but not aluminium on aluminium. For pressure-fed lubricated bearings one surface is often made of a lead alloy or, in aircraft engines, a silver alloy. The friction between surfaces can be reduced by introducing a low friction material or fluid between them. Grease, oil, graphite and even air are used for this purpose. Fluids have different qualities when moving, so the choice of a fluid as a lubricant is important. The ability of the layers of a fluid to slide over one another is referred to as viscosity.
A highly viscous fluid causes a noticeable resistance to movement if used as a lubricant. A fluid with a very low viscosity will not be a suitable lubricant either as it will not be able to keep the bearing surfaces from rubbing on one another unless fed into the bearing under considerable pressure.

1 Measuring the wind

Think of the effect of the wind on the work of such people as fire-fighters, sailors and navigators, civil engineers and architects. Wind measurement is a vital part of their work. In this section you are asked to make wind measuring instruments.

The Beaufort scale is a simple means of measuring the speed of the wind by comparing the effect of the wind on objects such as trees, waves, flags.

Light breeze
Force 0-3

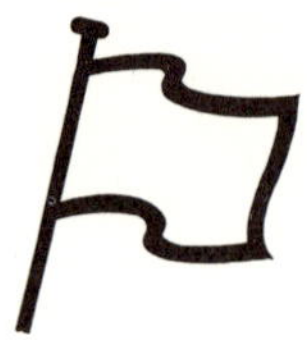

Moderate breeze
Force 4-5

Strong wind
Force 6-7

Many people need to measure wind speed more accurately than this. For example, in civil engineering, success, failure or even disaster depends on, amongst other factors, the assessment of wind effects. Models of large structures such as roofs, towers or bridges are tested in wind tunnels in order to simulate and measure the effect of the wind with great accuracy.

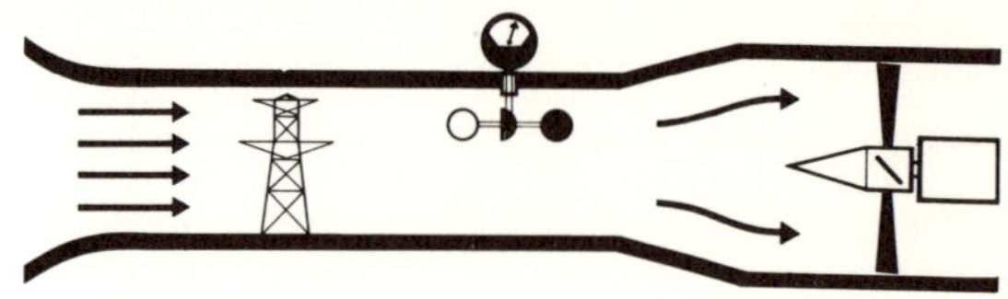

An anemometer is an instrument that measures the speed of the wind; a weather vane measures its direction. There are many ways of making these instruments.

INVESTIGATION

A REVOLVING VANE

Look at the diagrams and answer the following questions.

Does the shape and size of the vane make any difference?

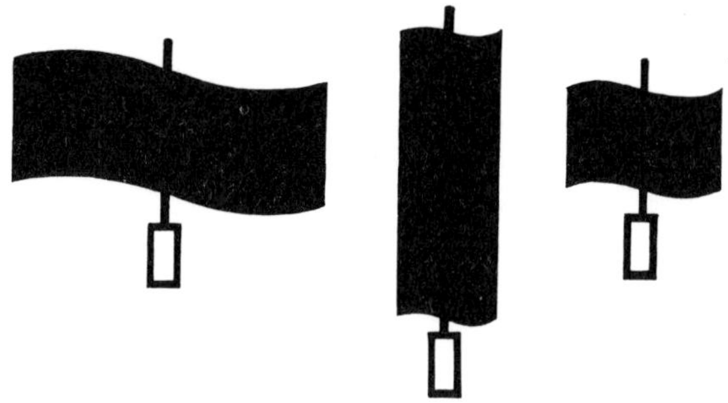

Is it possible to use a flat vane?

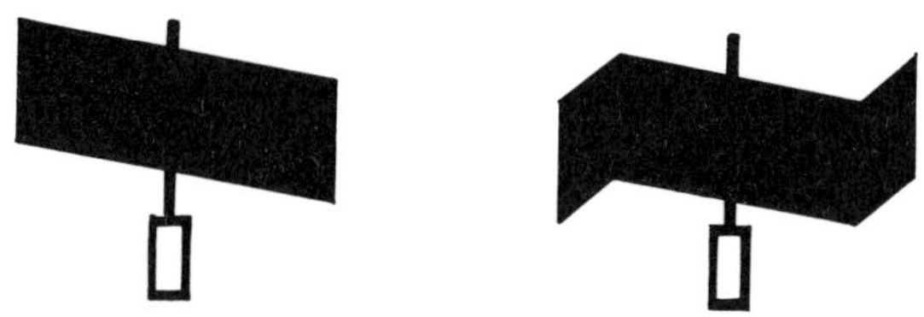

Would more than 2 vanes be better? If so, how many?

REVOLVING CUPS

Look at the diagrams and answer the following questions.
Does the size or shape of the cups affect the result?

Are long arms better than short?

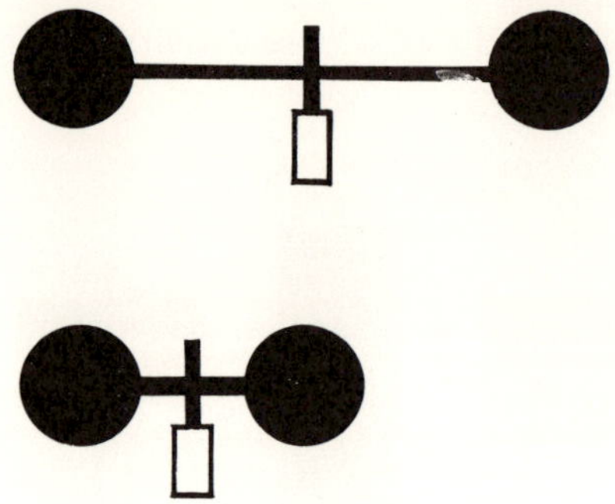

Could the cups be mounted with the axle horizontal?

Does the weight of the revolving parts matter?

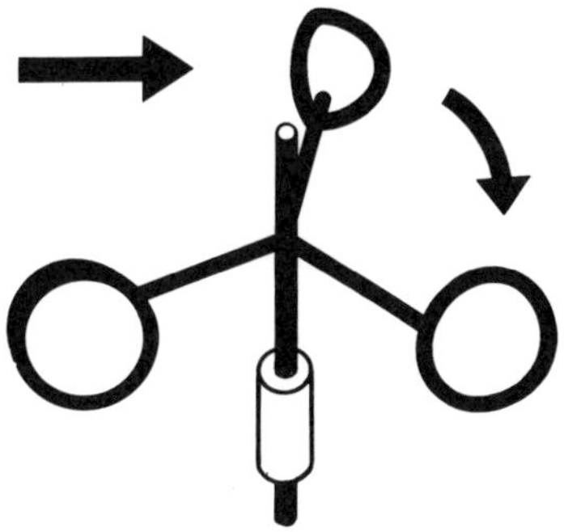

THE TURBINE

Look at the diagrams and answer the following questions.

Would a large diameter turbine blade work better than a smaller one?

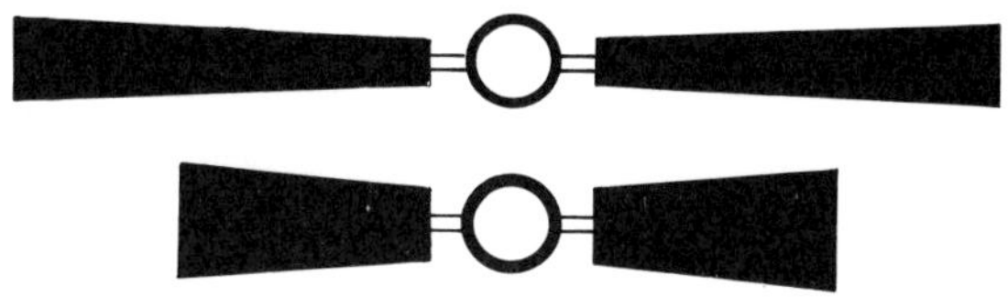

How many blades give the best result?

Does the pitch angle θ make any difference?

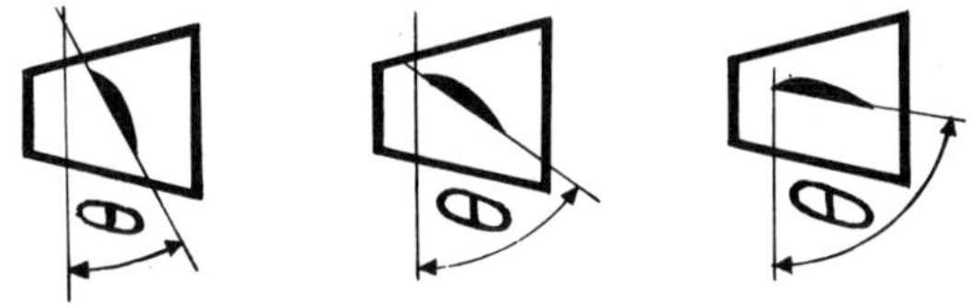

Which aerofoil section is best?

THE HINGED VANE

The stronger the force of the wind, the more the vane is blown from its vertical position.

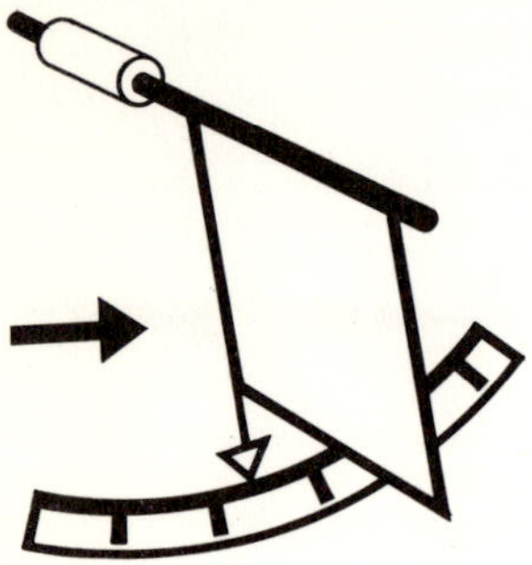

How could you make it read in very strong winds?

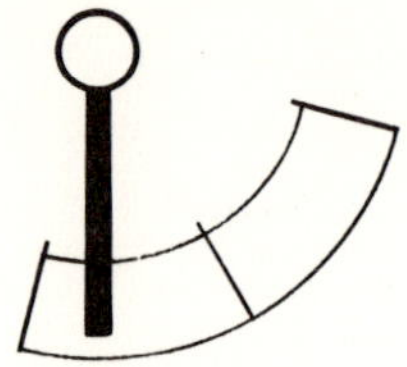

No wind

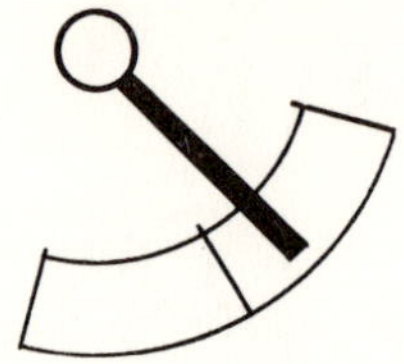

Moderate wind

Strong wind

How can you keep it facing into the wind?

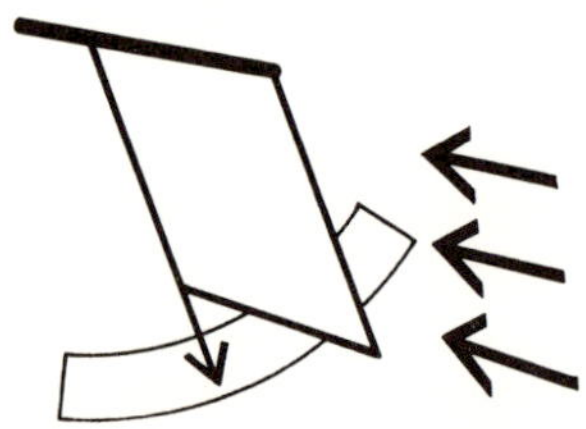

THE BELL-MOUTHED ANEMOMETER

This faces into the wind and is made from a U-tube partially filled with water. When making your anemometer consider the following.

Is another liquid more suitable?

How can you keep it pointing into the wind?

What is the relationship between head of water, air pressure and wind speed?

Can the instrument be made more sensitive?

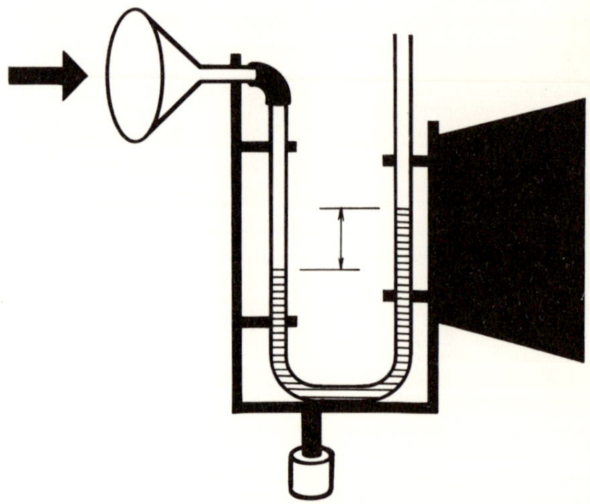

HOT WIRE ANEMOMETER

This is made from a torch bulb with the glass envelope carefully removed.

Being very compact and containing no mechanically moving parts, it is suitable for use in a small wind tunnel.

The air flow over the bulb cools it and causes a change in resistance. This is measured on a Wheatstone bridge, the bulb being one arm of the bridge, the other arms being high grade wire-wound resistors. Balancing is done using a null balance potentiometer of 1 000 ohms resistance. The measurements are made on a sensitive galvanometer. The cooling effect of the wind can be used to measure the speed of the wind. For example, the faster the wind, the greater the cooling effect, the less the resistance in the bulb wire and so more current flows across the bridge. The galvanometer measures the minute potential difference caused by the changing conditions.

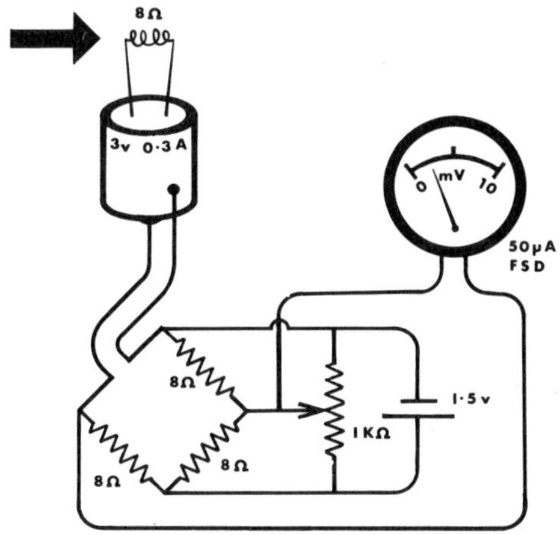

DESIGN GUIDE

When making your anemometer consider: what materials to use, how to shape them and fix them together; the strength and rigidity of parts and assemblies; how to make the working part weatherproof; rust, corrosion and maintenance; friction and wear.

How long will it take to make and can you easily obtain the materials you have selected?

What will it cost?

Models can be made using simple materials like cardboard, cork, wire, wood, plaster and plasticine.

Trials will help you to decide which type to make and help with the design of detail parts.

A meteorology station would benefit from having an automatic 24 hour record of information given by the anemometer.

WIND DIRECTION

What about the direction of the wind? Would a normal weather vane be suitable for your purpose?

Flying your model aeroplane

Sailing a boat

Recording weather conditions

You may wish to measure wind direction very accurately or by remote reading methods – so that the reading can be made at night in a room, and recorded on a revolving drum (see Appendix for further details).

2 Measuring weight

You are asked in this section to make a weighing machine. When deciding what you want it for, the following points should be considered.

What do you want the machine to weigh?
(a) Heavy objects (like a sack of poatoes).
(b) Medium size items (a basket of fruit).
(c) Light objects (letters).
How accurate need it be?
(a) 1 kg in 50 kg.
(b) Within 20 grams (+ or − 10 grams).
(c) Within 2 grams (+ or − 1 gram).

EXAMPLE

Design and make a machine for weighing live fish.

Because these fish are to be thrown back into the water immediately after weighing, the machine should fulfil the following conditions.
(a) Have a range of 0 to 20 kg.
(b) Weigh to an accuracy of + or − 2 grams.
(c) Fast acting.
(d) Have a pan rather than a hook.
(e) Portable (light and compact).
(f) Non-rusting
(g) Inexpensive.

Before you can select the most suitable type of machine for your purpose it is necessary to find out how different types of machine work.

Most machines either compare the load (object being weighed) with a standard mass called a 'weight' or they measure the effect gravity has on a mass by causing it to extend a spring of known strength and rate.

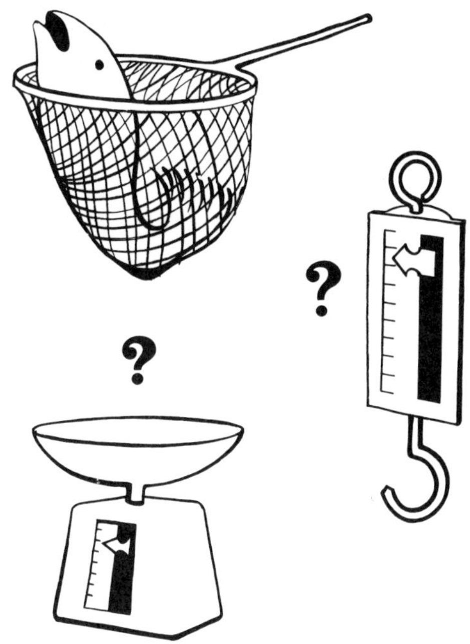

TYPES OF WEIGHING MACHINES

1. THE SPRING BALANCE

A scale indicates how much the weight has extended the spring.

Springs are made in many different lengths and diameters with varying numbers of turns and different sizes of wire.

The amount of extension caused by a certain size of weight is known as the 'rate' of the spring. If a 1 kg weight stretches the spring 20 mm then the rate is 20 mm per kg.

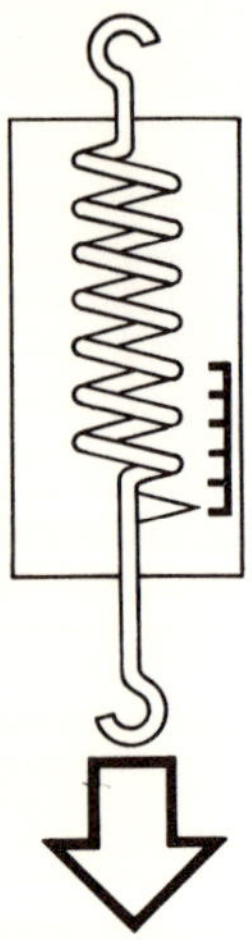

2. SCALES

This type of machine has been in use for many centuries. It uses a set of standard weights to balance the object being weighed by trial and error.

Very good results can be obtained with scales and according to its use they can be simple or elaborate. The pointer attached to the centre of the beam is essential.

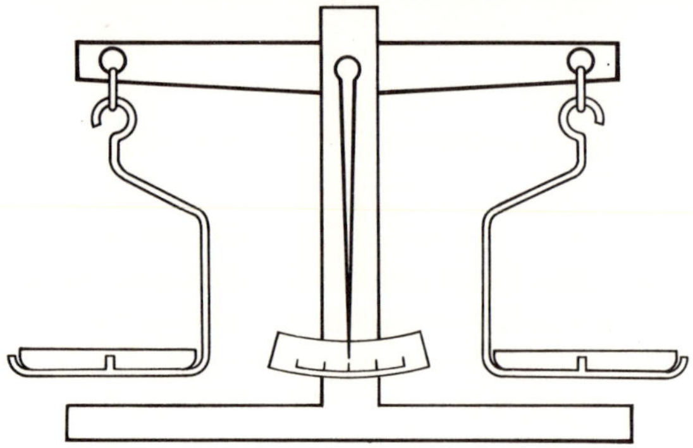

Some conclusions can only be reached after a practical investigation had been made of the basic principles and problems. You will need to decide, for example, the best type of centre pivot.

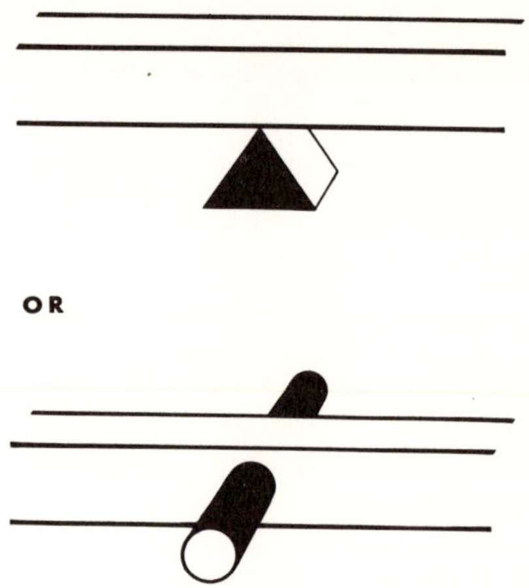

Consider the material and the shape.

How long should the beam arms be?

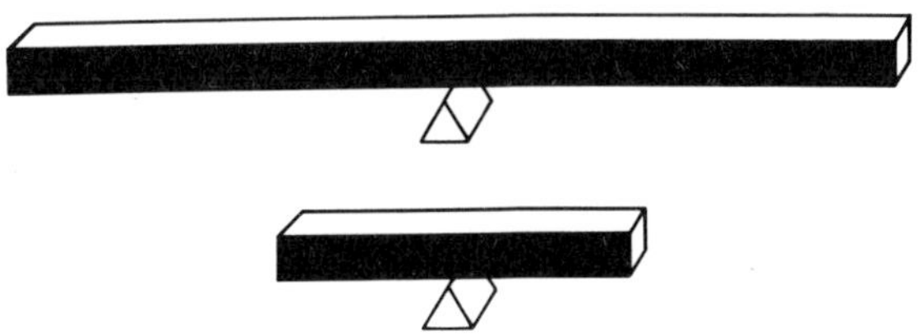

Is the weight of the arms important?

What type of pivot will you use for the beam ends?

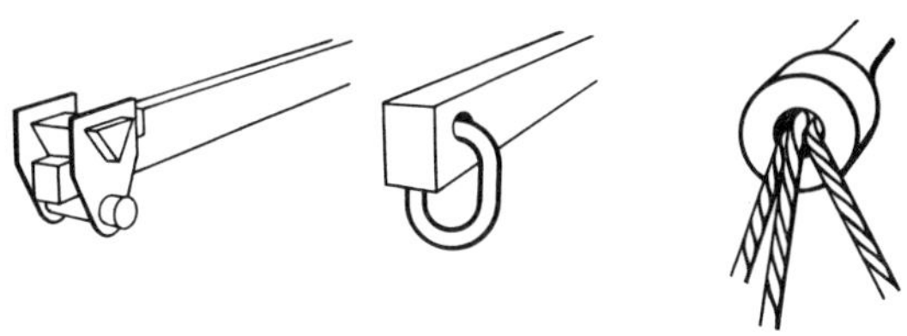

Knife edge | Hole and ring | Cord and trumpet end

3. STEELYARD

In this type of scale the load is hung at a different distance from the pivot than the 'weight'. The weighing is done by moving the weight until it balances the load. By varying the distance the need for having a selection of different weights is avoided.

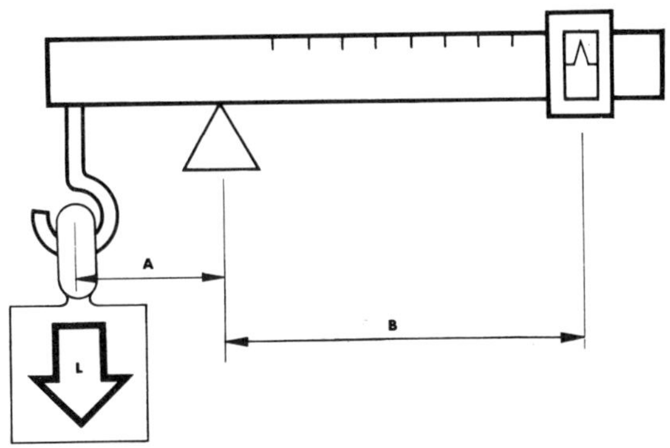

EXAMPLE

If a load (L) 100 mm from the pivot is balanced by 1 kg mass (the weight) 500 mm from the pivot then $L \times a = W \times b$. If W, a and b are known then L can be calculated. The four steps in the calculation are:

(a) $L \times 100 = W \times 500$,
(b) $L \times 100 = 1 \times 500$,
(c) $100\ L = 500$,
(d) $L = 5$ kg.

4. THE COUNTERPOISE

This machine is designed to work without the use of additional weights.

A heavy load (L1) pushes the link mechanism so that the counter-weight on the pointer (W1) is moved nearer to the horizontal than it is with a lighter load (W2). This machine relies on the principle that the moment of a force is the force multiplied by its perpendicular distance from the pivot. The effective distance of W1 to the pivot is greater than W2. This is a convenient, robust and accurate type of machine, often used by shop keepers. Its sensitivity depends on the pivoted joints being as frictionless as possible.

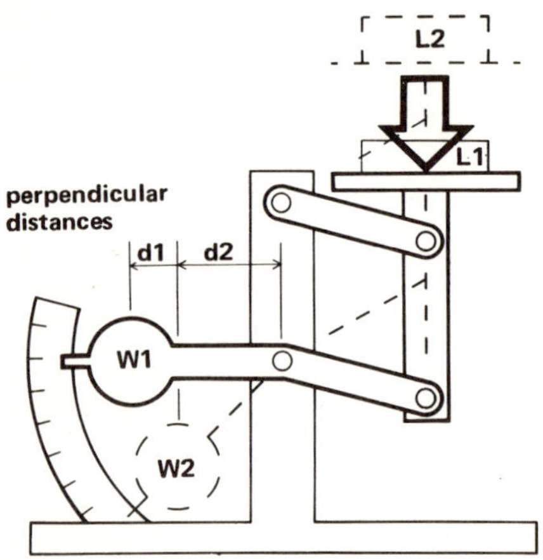

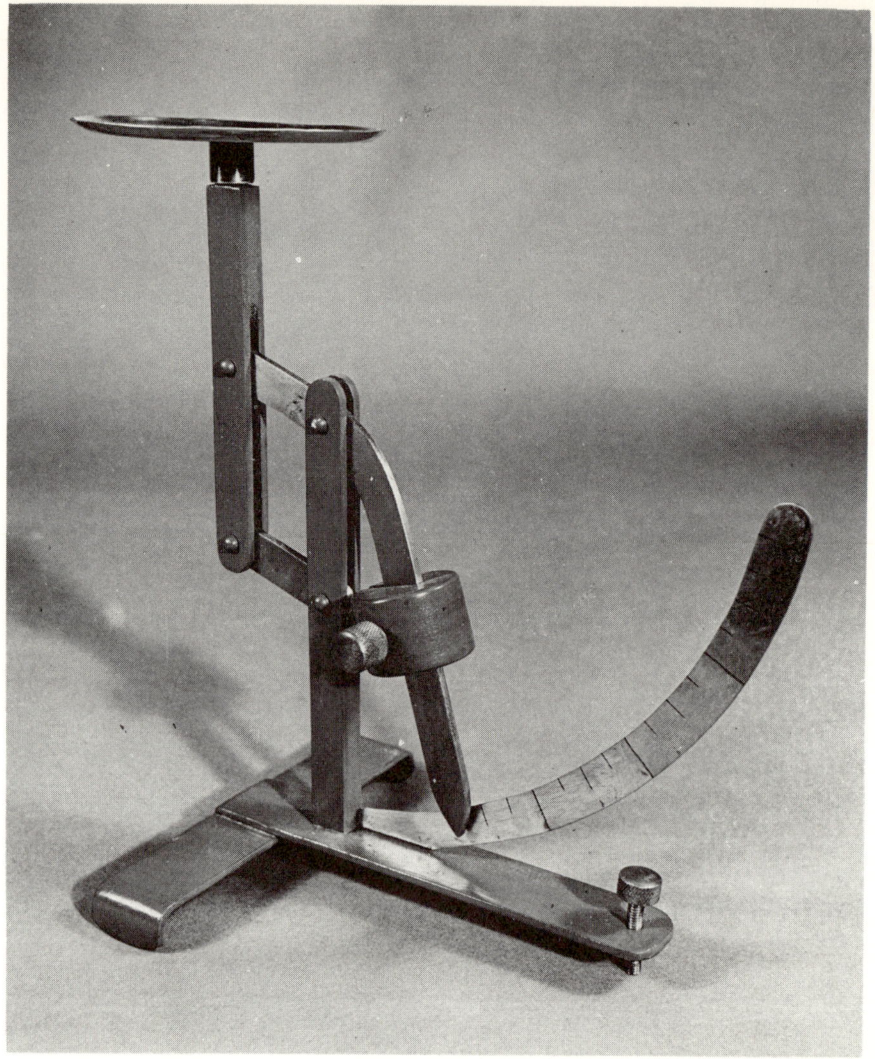

When your investigation of all the types has led you to choose a particular machine you will have to consider in detail many other problems, such as:

The method of suspending the pans from the machine — wire, strip, cord?

The materials and methods of making pans of the same weight (metal sheet, spun, beaten or pressed into shape; or plastic, cast, blown or vacuum formed into shape).

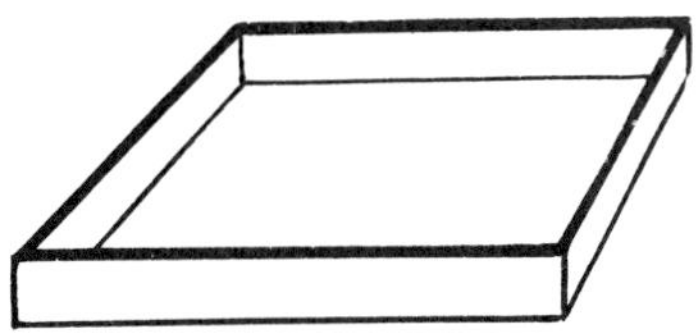

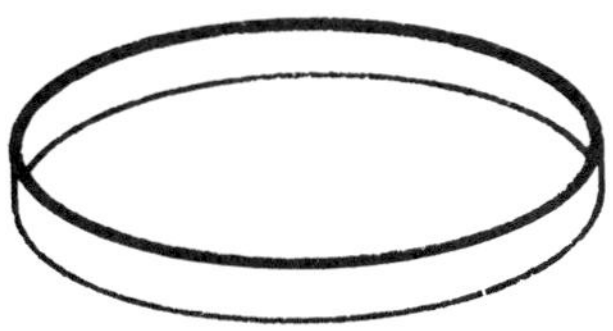

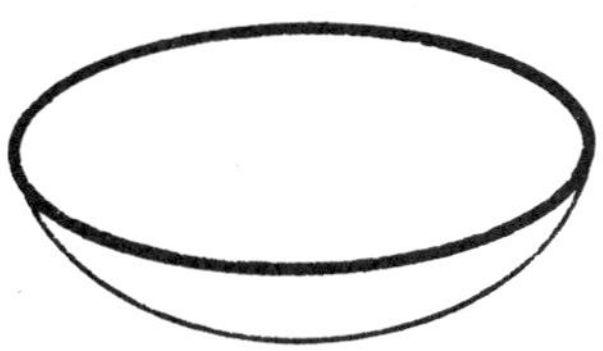

The base must keep its shape and has to be level. What materials are suitable and what device can be used to indicate the accuracy of levelling?

Unless some form of damper is incorporated some machines will see-saw for a long period before coming to rest.

Most machines will require some form of adjustment to compensate for very slight differences between the weight of the arms, pans, etc.

If the machine is allowed to corrode, its accuracy and appearance will be ruined. Working parts such as bearings, arms, pans, should not be painted. Suitable methods of protecting the machine against corrosion, dust and, in some cases, vibration will have to be devised.

No matter which machine you make, friction, lightness of moving parts, methods of indicating the weight, levelling and protection against corrosion must be given great attention.

DESIGN GUIDE

Suppose the spring balance is selected for weighing the live fish before they are thrown back in the water (page 35). It fits the requirements most closely, particularly in being quick acting, portable and inexpensive.

In order to select a suitable spring some tests are necessary.

HOW TO START

Make some springs by winding annealed high carbon steel (such as piano wire) on to a mandrel (wood or metal cylinder) held in the lathe between chuck and centre.

The pitch or distance between the coils should be constant and after the ends have been formed the spring must be hardened and tempered. Two main facts will affect the

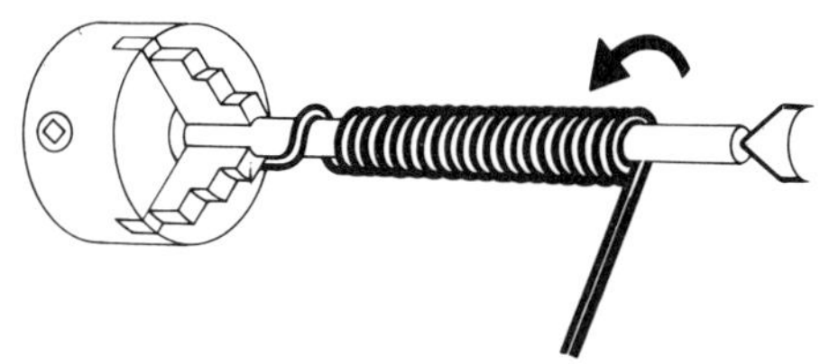

strength of your spring; these are the diameter of the wire and the diameter of the spring. Springs wound on the same mandrel but from different gauges of wire will be of different strengths. Similarly springs made from the same diameter wire but wound on different mandrels differ in strength. Investigate the performance of each spring by making a device similar to the one shown (right). This one is for testing springs in tension.

When you have made various tests between different springs and between springs in tension and in compression you will select a system and a spring which suits your requirements.

The compression spring device will be more difficult to make free of friction than the balance which uses a spring in tension.

It is possible to use two or more springs at once. Sometimes this helps to overcome problems of heavy loads and friction.

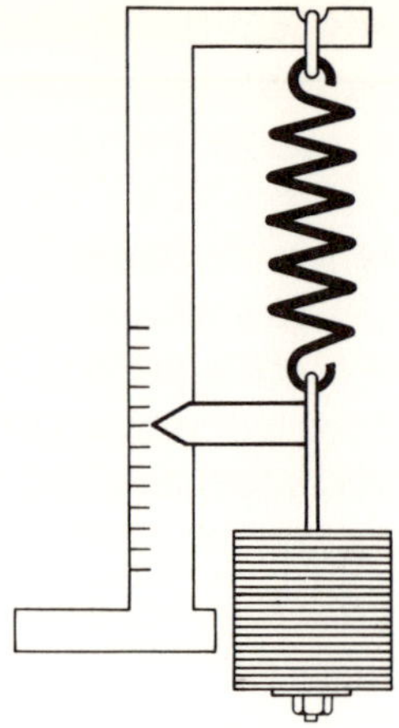

CONSTRUCTION

The machine shown in the photo opposite has two tension springs in it; the weight is indicated by a pointer driven by a rack and pinion. Such systems are hard to make, particularly if they are to be free of friction and backlash.

There are simpler alternatives which give good results. (See the sections on Mechanisms and Barometers.)

The case is made from mild steel, the pan from brass and the base from a hardwood.

The position of the pointer on the spindle is adjustable, allowing it to be brought to zero when necessary.

Most of the machine, except the working parts, is painted with hammer finish paint.

MORE ADVANCED METHODS OF WEIGHING

BY THE MEASUREMENT OF THE DEFLECTION OF A BEAM

A beam bends slightly under a load. If not over-loaded the beam will return to its original position when the load is removed.

A dial gauge can be used to measure the deflection for a series of known loads, then calibrated for all loads.

Find out the best materials for the beam. Investigate other methods of measuring the strain in the beam.

Two further suggestions - use a photo-elastic material in a polariscope to show the strain patterns. Look into the use of strain gauges and load cells.

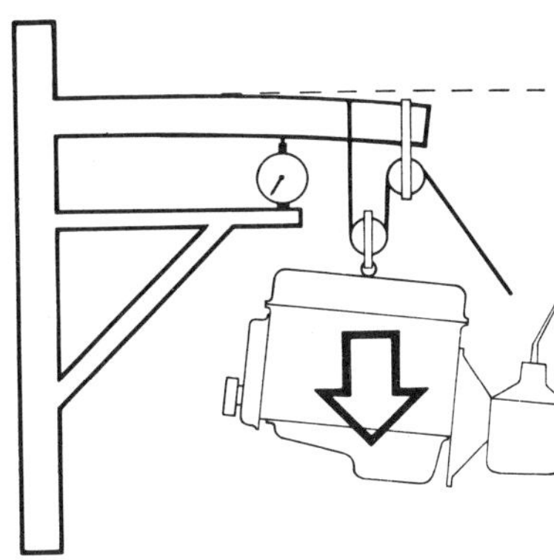

BY HYDRAULICS

A hydraulic cylinder (like a brake cylinder on a car) which is attached to a pressure gauge will form the basis of a weighing device. A suitable hydraulic fluid is required together with piston seals which will not allow leakage.

This machine will not be very sensitive but if suitably made, will take very heavy loads. How do you work out the size of cylinder to make it suit a particular size of gauge?

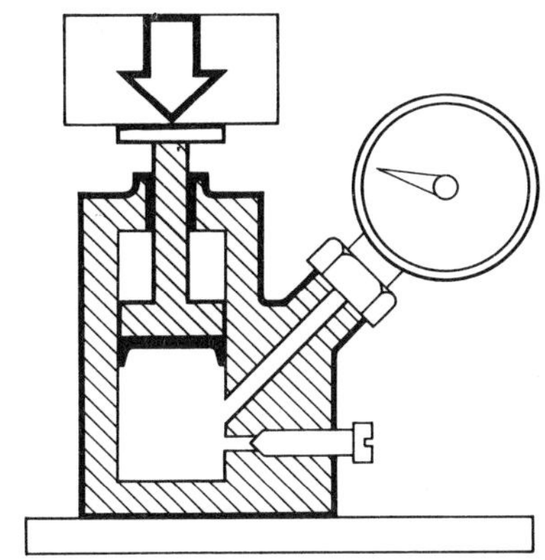

COUNTERPOISE MACHINES WITH A LINEAR SCALE

You will find that on a simple machine (see photo on page 38) the graduations on the dial will be unequally spaced. Can you devise a way of making the pointer advance the same distance each time an additional load of the same size is added to the scale pan?

The machine will take some seconds to come to rest after a load has been placed on it. This wastes time. How can the mechanism be damped to make the pointer come to rest quickly without affecting the sensitivity of the machine?

UNUSUAL WEIGHING

Ships are weighed by displacement. That is, by measuring the volume of water pushed out of the way by the ship as it floats in the water. When floating, a ship has displaced exactly the same weight of water as its own weight.

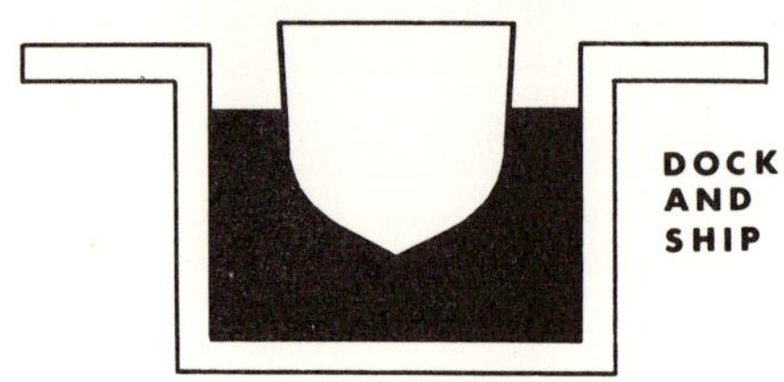

To avoid capsizing, boats are often loaded with ballast to lower them to the correct level in the water. If you have a model boat, particularly a submarine, try to invent an 'upside down' weighing machine to measure the amount of ballast needed. Remember, the machine must measure the force required to push the ship down in the water to the correct level.

3 Measuring temperature

Thermometers are used for measuring temperature and are made in many different forms according to their use. Make a thermometer, using the details in this section for guidance.

Here are three situations requiring temperature measurement.

1. Body temperature

2. Car engine temperature

3. Kiln temperature

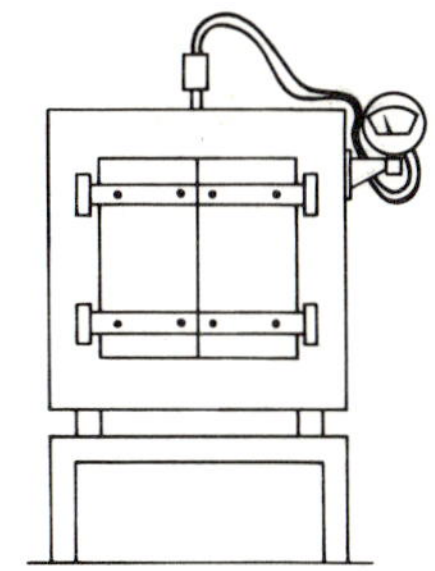

For each particular situation a list of requirements must be made which will enable a brief to be drawn up. Here is an example.

A CLINICAL THERMOMETER

(a) Range: 35° to 45° C.
(b) Graduations: in 0.1^{o} divisions.
(c) Size and weight: such that it can be easily held in the mouth.
(d) Reading: visual, which remains until reset.
(e) Non-poisonous and easily cleaned.
(f) Cheap in price.

INVESTIGATION

Discuss *2* and *3* then write out a list of requirements for thermometers for each situation. Remember to include in your list the following points.

(a) The range ($-20°$ to $0°$, or $0°$ to $100°$, or $0°$ to $200°$, or $200°$ to $500°$C for example).
(b) The required graduations (every $0.01°$ or $0.1°$, or $1°$ or $10°$, or $50°$C etc).
(c) Size and weight (depending on situations and use).
(d) The method of reading (direct and visual, or visual reading on a scale situated some distance from the heat source, etc.).
(e) Any other features (robust, non-poisonous, protected against impact, etc.).

Danger

Before using any chemicals or materials in situations new to you, be careful to find out whether they are inflammable, explosive, poisonous and at what temperatures they boil and freeze.

TYPES OF THERMOMETER

1. A LIQUID IN A SEALED TUBE

Liquids expand when heated. If a liquid is contained in a glass bulb which has a single opening into a narrow tube, the expansion is shown by the liquid being forced into the tube. The amount by which it expands, and the length of tube that consequently fills, depend on the rise in the bulb's temperature.

Some liquids are unsuitable for use in this kind of thermometer because they boil or freeze within the range of temperatures the thermometer may be used to measure. Mercury or alcohol are widely used.

Note: Mercury is poisonous. Alcohol boils at $78°$C. Glass is brittle.

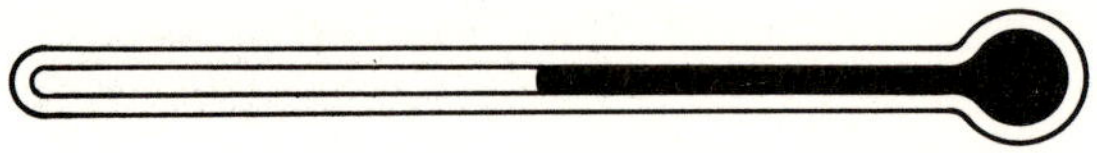

Answer these questions
How is the bulb made, filled and sealed?
How is alcohol made visible?

2. A BIMETALLIC STRIP

A bimetallic strip is made from two layers bonded together, each of a different kind of metal. As the strip is heated, one metal expands more rapidly than the other. This causes the strip as a whole to bend. (Can you see why?). On cooling, the strip returns to its original shape.

The drawing shows a bimetallic strip being used to close the electrical contacts in a bell circuit. When the temperature reaches a certain level the strip bends upwards sufficiently to close the circuit.

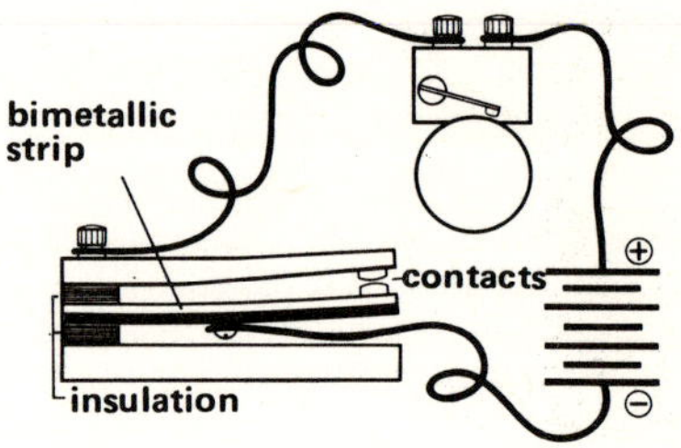

Answer these questions
Which metals are suitable?
How long, thick and wide should they be?
How can they be fixed together?
Could the device be adjusted to ring a bell at a required temperature?

3. A VAPOUR PRESSURE THERMOMETER

As the vapour pressure from a liquid increases due to a rise in temperature the reading on the pressure gauge rises.

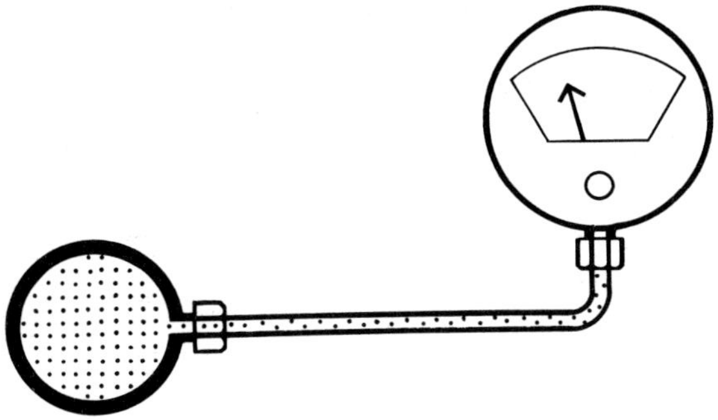

Answer these questions

Which liquid is suitable?
How can the bulb be filled?
How can the tube be fixed to the gauge and to the bulb?

DESIGN GUIDE

Suppose your choice is to construct a bimetallic strip thermometer. It might be used either as a room temperature thermometer or as a thermostat for controlling the temperature in an animal house.

Two or more shapes of bimetallic strip could be used - spiral or flat being examples.

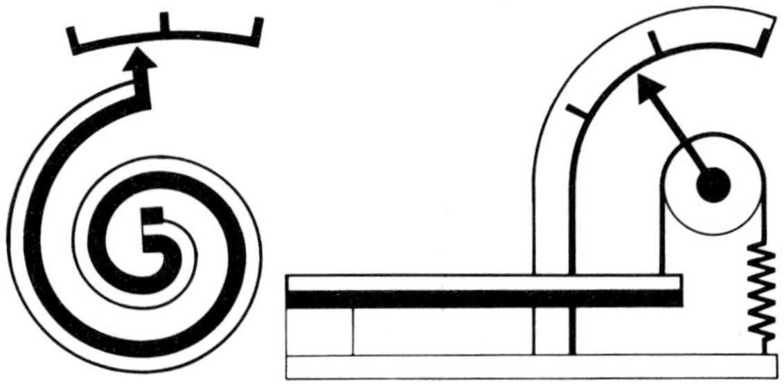

DISCUSS

(a) Which two metals are suitable for making the strip? (They should be easily obtained, cheap and easy to work.)
(b) How can the strips be fastened together? The spiral one will cause special difficulties.
(c) How can you obtain various known temperatures for experimental work?

INVESTIGATE

1. The success of soldering, bonding and riveting the two strips together.
2. The movement of different lengths of bimetallic strip over the same range in temperature by fixing one end, heating and measuring it.
3. Whether a wide strip is better than a narrow one.
4. Methods of attaching the strips to a pointer.

Below is a simple bimetallic strip thermometer made to measure room temperature. The strip was made from aluminium and steel (tin plate) riveted together.

Your thermometer could be much more compact than this one.

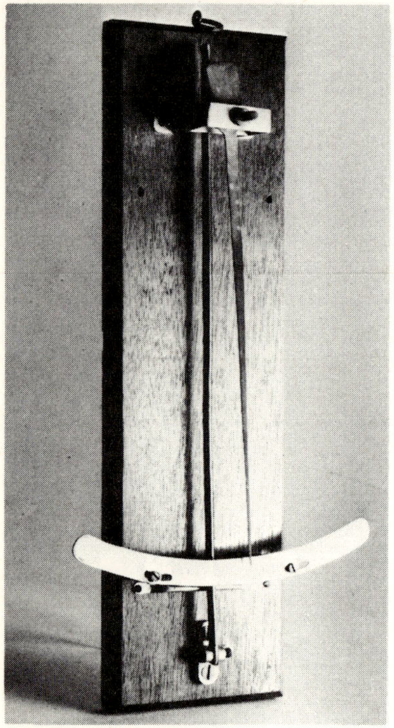

OTHER SUGGESTIONS FOR MAKING USEFUL THERMOMETERS

1. A THERMOCOUPLE

This thermocouple could be used for a domestic oven, kiln, or furnace, or for measuring the temperature of metals melted ready for casting.

The hot junction is made from two different metals twisted or welded together.

When the joint in the wires is heated a very small electric current is produced (the thermoelectric effect). The greater the heat applied to this joint (hot junction) the more electricity is produced. Some couples produce their greatest electric current at a higher temperature than others.

For 0-400°C use a copper/iron junction.
For 0-1350°C use chromel/alumel (known as T1, T2).

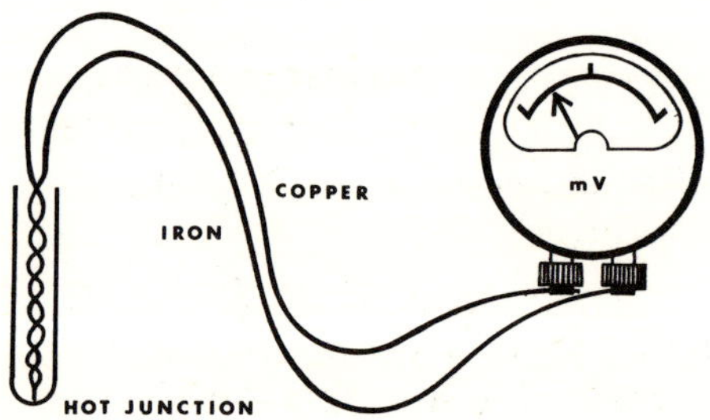

2. AN AUTOMATIC RADIATOR SHUTTER

These are the things you want it to do: to open at a radiator water temperature of 70°C; to 'fail safe' and have a mechanical override; to be powerful enough to overcome the rush of air through the radiator grill; to be rust-proof.

Do not use an inflammable chemical in the bellows.

Note: 'fail safe' means that if the mechanisms fail, they do so with complete safety.

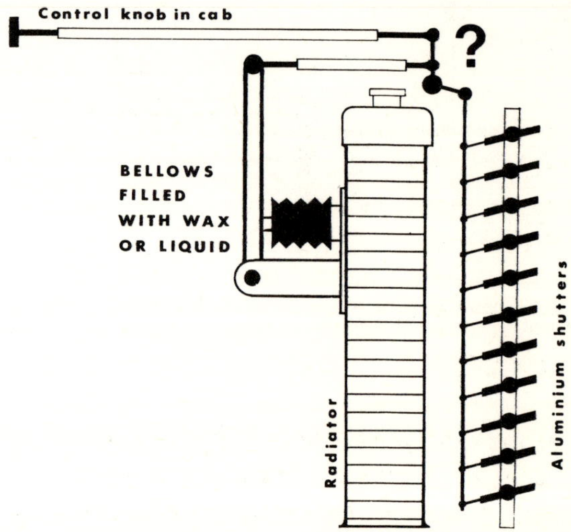

3. THERMOSTAT FOR AN INCUBATOR

You want it to turn a heater on or off; to 'fail safe' to be completely reliable and require little maintenance; to be cheap to make and run. The container and the bulb should be in the incubator.

Why is dry air necessary in the container?

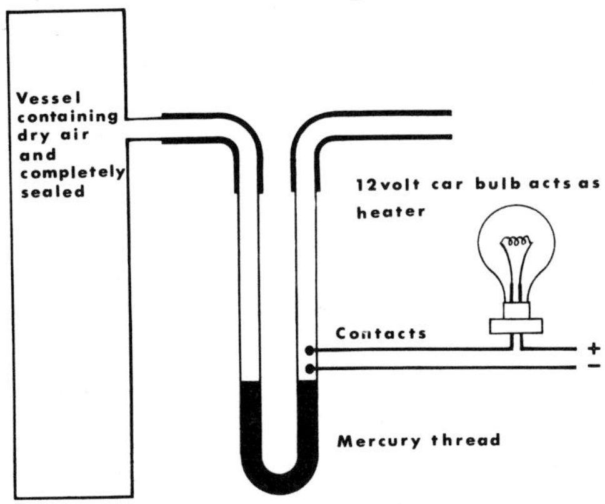

EXAMPLE

Suppose your choice is to make a pyrometer for foundry work and you have chosen to measure the temperature of molten metal with a thermocouple.

DESIGN GUIDE

1. Range 0° to $800^{o}C$.
2. Graduations every $20^{o}C$.
3. *Size* - 2 metres long to allow the junction to be immersed in the crucible of molten metal; *weight* — for ease of handling not more than 3 kg.
4. Reading — on a scale at the top end of the supporting tube.
5. To be robust but uncomplicated.
6. To cost as little as possible.

You will need the following. A millivolt meter, preferably 100 mV fsd (a less sensitive one will give less accurate results). A steel tube for providing support. Some thermocouple wire for making the hot junction (copper/iron, copper/constantan, chromel/alumel).

Find out before making

1. What readings are obtained from the different junctions when placed in molten lead, molten aluminium and the hot part of a bunsen flame?
2. How the junction can be protected to lengthen its life and increase the range of the instrument. Try shielding it with various materials which will withstand immersion in molten aluminium.
3. Which is the best way of fixing the meter and the junction to the supporting tube?

Danger

Cold metal immersed in molten metal can cause an explosion.

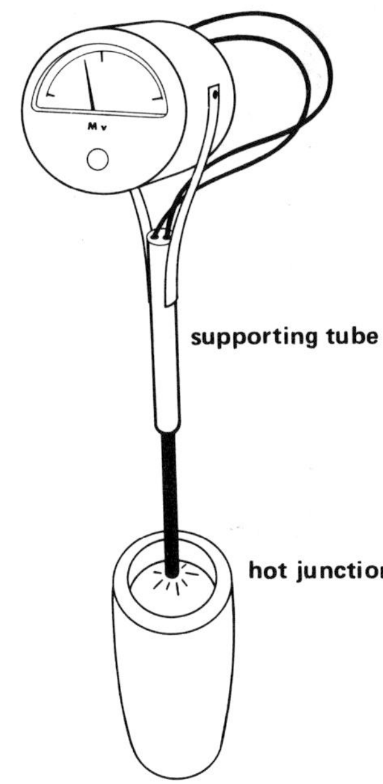

4 Measuring air pressure

A rocket on its way into space has to travel upwards through a layer of air approximately 100 miles deep which presses down upon the earth's surface. This pressure can be measured accurately with a barometer, but its effects can be seen in the experiment below. Apply the principles and techniques in this section to make a barometer.

INVESTIGATION

Investigate air pressure in these ways.

1. Turn a milk bottle full of water upside down into a bowl of water.
As long as the bottle neck is under water, no water will run out. Why not?

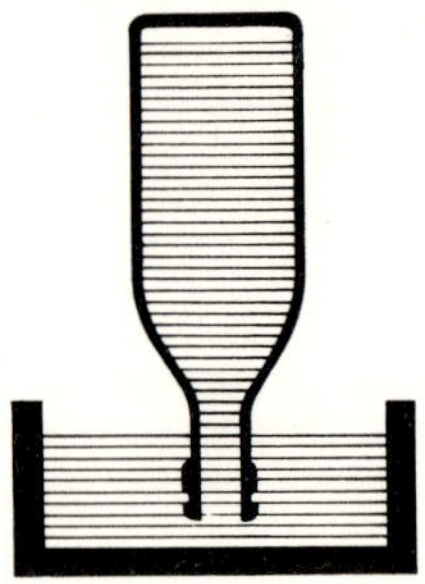

2. Does the same principle hold true in the case of a larger mass of water?

Try the experiment with a large container in a pond or bath.

3. If you connect a pump (or the inlet of a brazing hearth compressor) to an airtight container you can pump out some of the air. If the walls of the container are weak it will collapse.

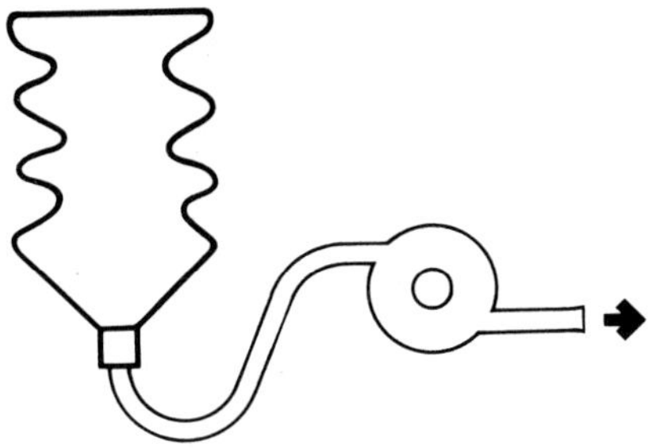

If the container is strong (like a fire extinguisher case) it will withstand the pumping (or exhausting) operation.

Your pump, unless it is a very expensive vacuum pump, will only be able to pump out part of the air.

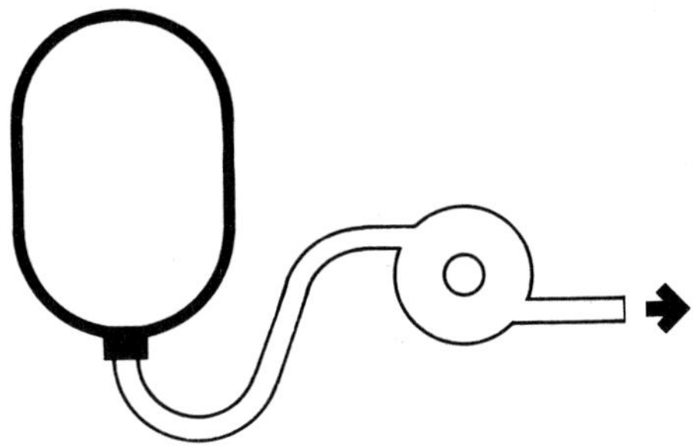

4. If you fill a tin can with steam, then put the lid on, it will collapse when the steam is condensed into water.

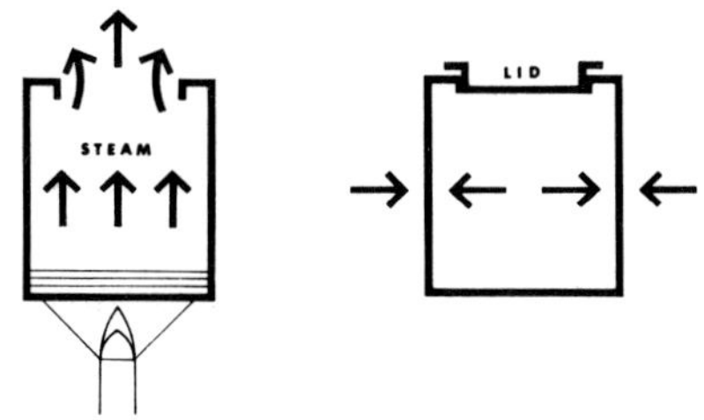

The air pressure outside the tin is now much greater than the air pressure inside. The thin walls are pushed in by the outside air.

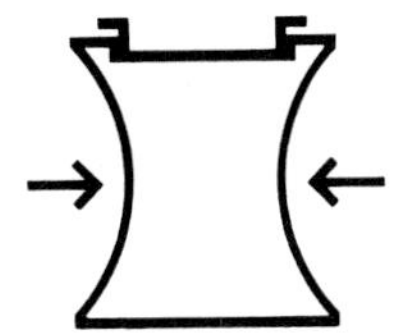

DISCUSS

1. The connection between FORCE, PRESSURE and AREA.
2. The units used to measure (a) atmospheric air pressure, (b) tyre pressures.

ACTIVITIES

Sketch Newcomen's atmospheric engine. Briefly say how it works.
Make a chart of your local atmospheric air pressure values (taken from newspaper charts or direct from a barometer). The readings will be in millibars and should be recorded daily.

We have seen how the weight of air above the earth will compress the air on the surface. The same principle applies to liquids. The weight of water in the sea causes the pressure of water under the surface to become greater according to the depth. A submarine can only go down a limited distance because of the danger of being crushed by water pressure.

INVESTIGATION

Water is many times heavier than air. A height of approximately 10 metres of water will produce the same pressure as the height of the earth's atmosphere pressing down on the earth at sea level.

You can prove this by fixing a pressure gauge at the bottom end of a hosepipe 10 metres high and filled with water.

If a transparent tube is used to seal the top of the hosepipe, and the bottom placed in water, the pipe can be lifted gradually until a space appears at the top of the tube (the space is a vacuum).

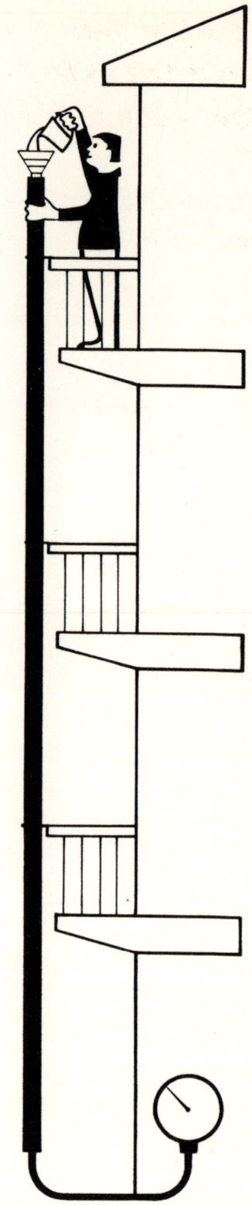

The distance from the surface of the water in the bowl to the level in the glass tube at the top is the exact height of water balanced by the weight of air in the atmosphere.

The height at which the weight of air supports a heavier liquid, such as mercury, is less.

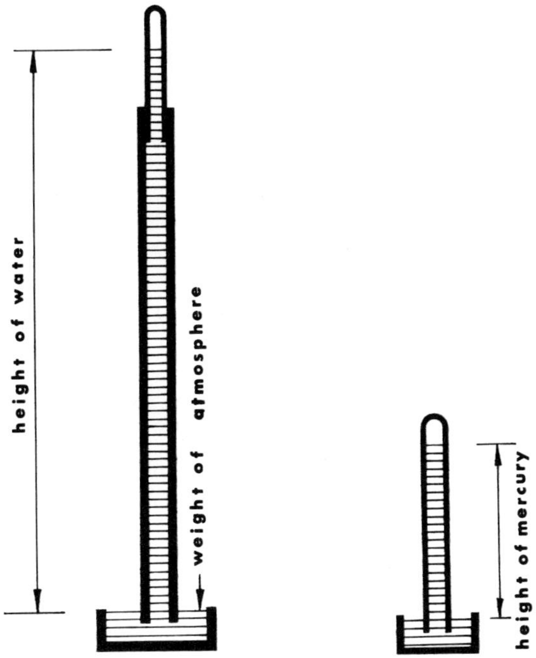

Liquid barometers are made with mercury because it requires a much shorter tube than water. Unfortunately mercury is expensive.

Barometers measure the slight differences in air pressure due to changing atmospheric conditions. A high air pressure can mean good weather, a low pressure can mean poor conditions. When the barometer 'falls' it indicates a lessening of air pressure.

TYPES OF BAROMETER

There are two types of barometer. Choose which you wish to make.

1. The liquid type.
2. The aneroid (non-liquid) type.

DESIGN GUIDE

Suppose you have chosen the aneroid type. It works on the principle of the crushed tin.

HOW TO START

A steel or brass container should be turned with a thick base and walls. A thin steel or brass disc (0.2 mm thick) is pressed out on a simple former so it is corrugated. The disc is fixed to the top of the container and a copper tube into the side, with soft solder.

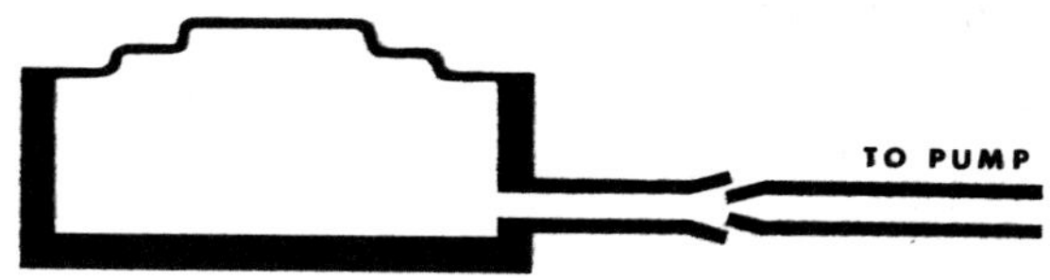

INVESTIGATION

How can you exhaust the container (capsule), (a) with a pump, (b) without a pump?
How can you seal the capsule, (a) temporarily, (b) permanently?
What materials are most suitable for the parts?
What method of showing the in and out movement of the thin wall of the capsule is best for your purpose?

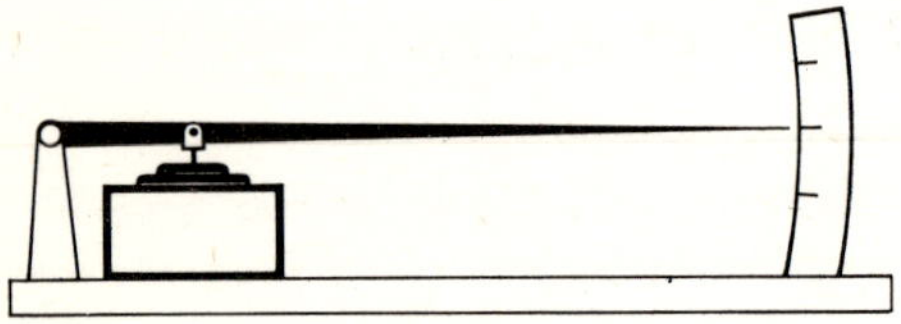

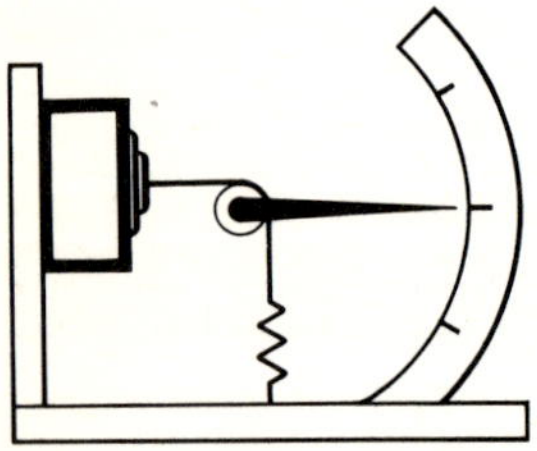

MORE ADVANCED CONSTRUCTIONS

A more powerful and sensitive instrument can be made with a capsule which has a thin base in addition to a thin top. Here is a section through such a capsule.

A simple press tool will enable corrugated capsules to be made.

The corrugation prevents the complete collapse of the chamber under high pressure conditions.

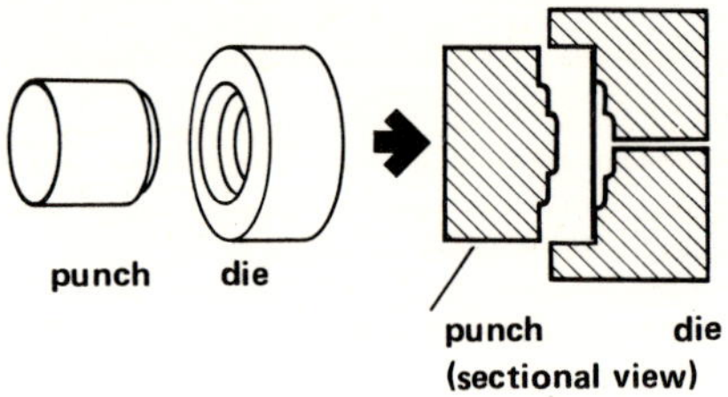

APPLICATION

A bank of capsules can be used to make a barograph.

The amount of movement is found by trial, and the length of arm designed accordingly. A drum is driven by a clock through two revolutions a day. Extra gears will have to be incorporated to obtain one revolution a week.

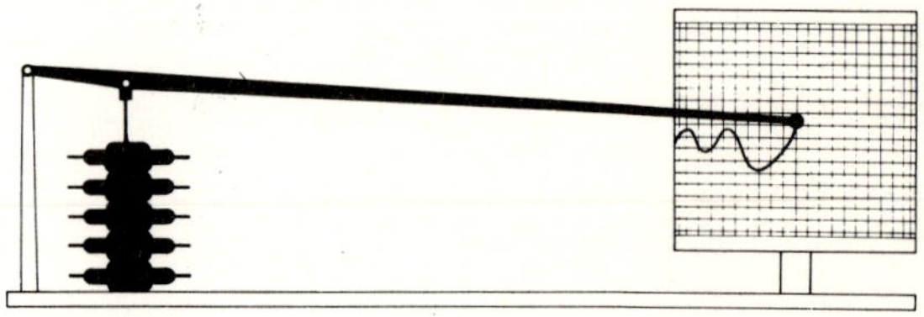

5 Pumps

You are asked to make a pump. This section gives you details which will help you to choose a suitable, interesting type.

If you wish to make a water pump for a particular purpose, such as for pumping water from a pond to a higher level or to run a garden fountain, then you must think first about whether you need: (a) a high pressure supply or (b) a low pressure supply, (c) a large volume of water or (d) a small volume of water.

1

Large volume
High pressure

2

Large volume
Low pressure

3

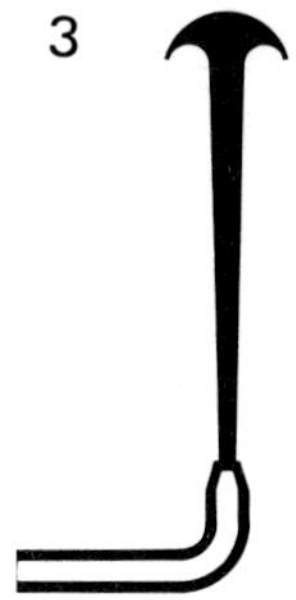

Small volume
High pressure

4

Small volume
Low pressure

The design of your pump will depend on the purpose for which it is intended.

See if you can decide what type of pump is needed to fulfil the requirements of the following examples. Add factors to your list (or brief) of requirements which you believe to be of importance. Safe, quiet working, cheap, expendable, reliable, not affected by corrosion, are some of many factors you may consider important.

1. To pump water for an artificial stream to find how life in a stream is affected by the movement of water, water purity, light and shade.

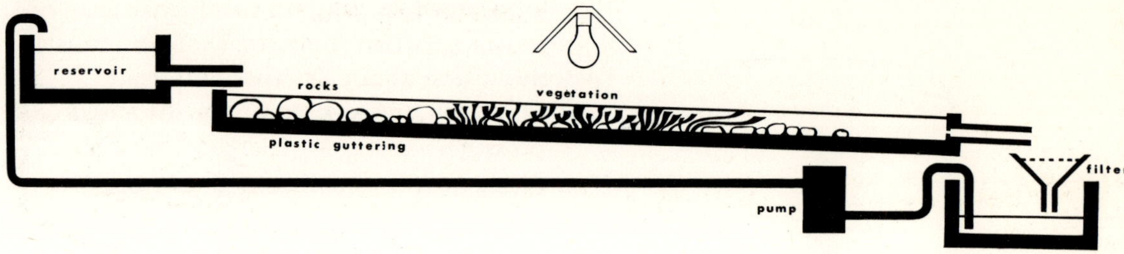

2. To pump water from a stream up to the roof tank in a cottage. How can you prevent over-filling the tank?

3. Feeding a car engine with petrol.

Note Petrol dissolves oil and rubber, evaporates easily and explodes when mixed with air and ignited by a spark.

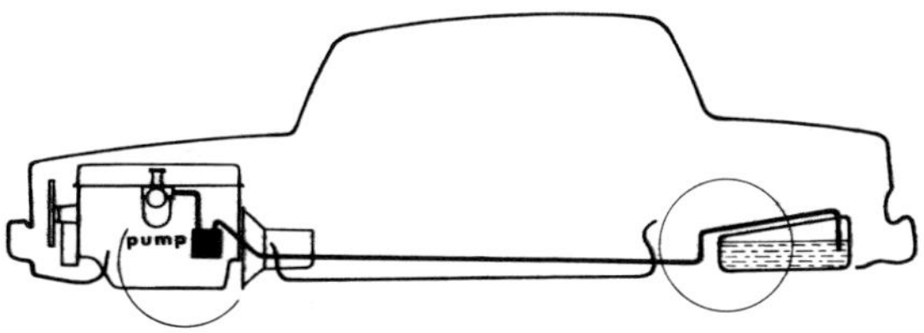

4. Running a garden fountain.

Note The use of electricity of mains voltage (240 V) is dangerous. Why?

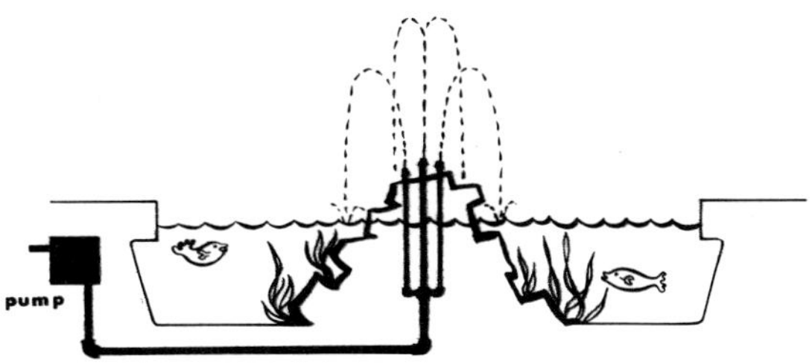

5. A water garden.

Note The water wheel would not be able to provide enough power for the pump to keep the circulation of water flowing continuously. Why?

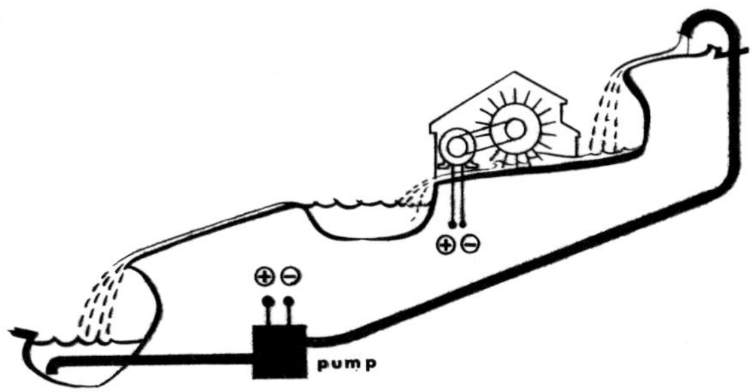

6. Pumping water through the cooling system of a car.

Note It would be useful to be able to check from inside the car that the pump was working.

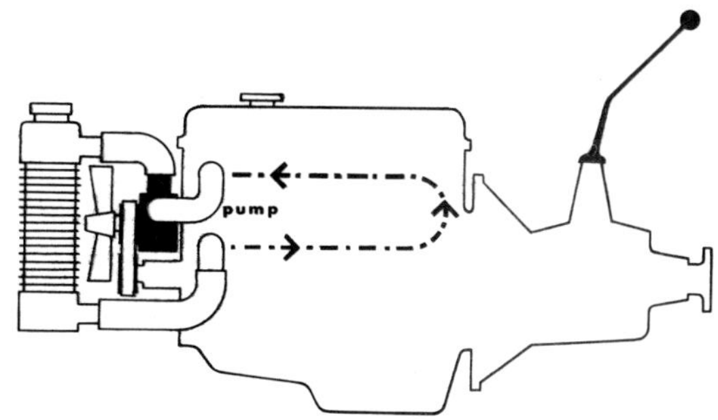

7. Pump for fire fighting.

TYPES OF PUMP

1. The piston and cylinder.

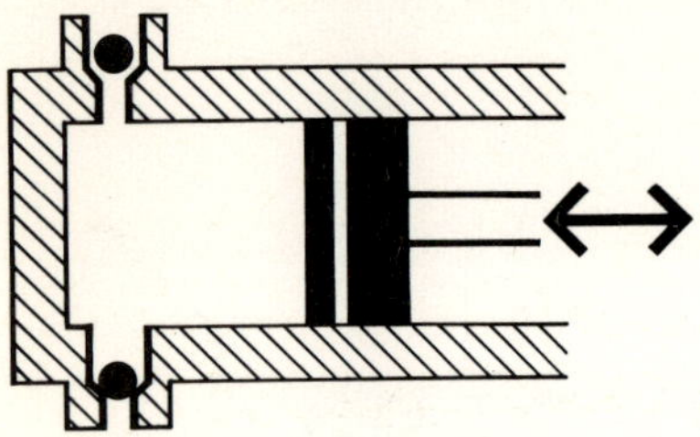

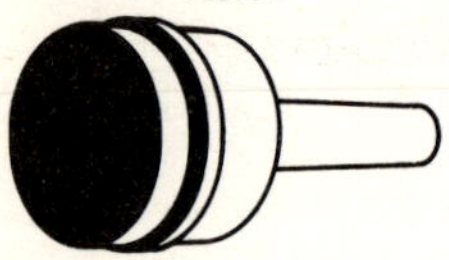

2. The diaphragm pump.

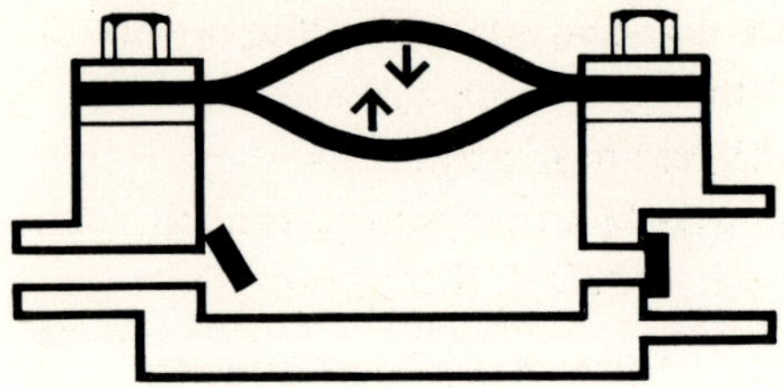

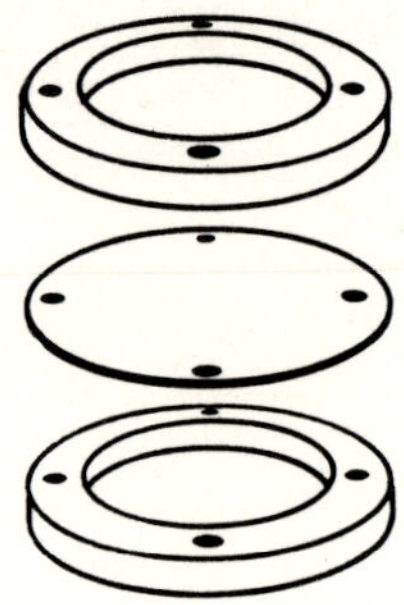

3. The centrifugal pump.

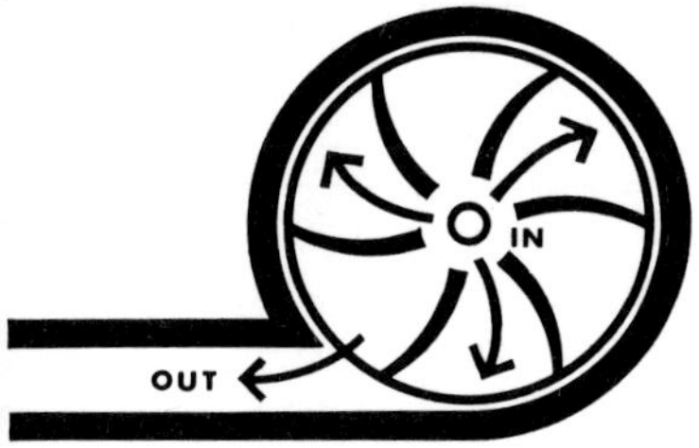

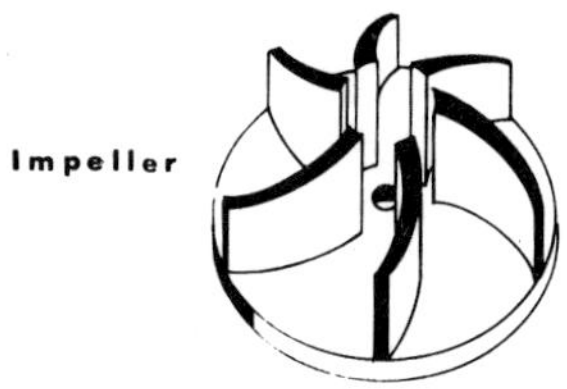

There are other forms of pump, some of which are extremely simple but ingenious. If (like the oesophagus) a flexible tube is filled with a liquid and squeezed, the squeezing action will move the liquid along the tube.

Can you invent a mechanical device which would simulate this action?

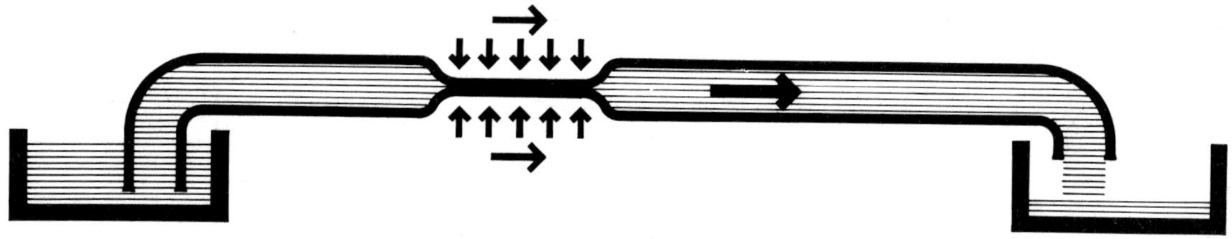

ANCIENT PUMPS

From early times men have used a variety of pumping devices. Look at these pictures of an Archimedian screw and a rag and chain pump.

Find out what they were used for and how they worked. Modern pumps are more efficient. Why?

What other ancient pumps can you discover? Make a list and sketch an example of each.

The Archimedian spiral screw pump. *Courtesy Ronan Picture Library and E. P. Goldschmidt & Co Ltd.*

A rag and chain pump in use in the 16th century.

INVESTIGATION

You can find out more exactly about the design of pumps by making some prototypes out of plastic, rubber, glass and metal tubes.

See what happens if you force a rubber piston up and down a glass tube which has its lower end immersed in water.

The 'cylinder' fills and empties as the piston moves up and down. To make the water travel outside the cylinder, tubes are needed, and inside these tubes there must be one way flow devices - valves.

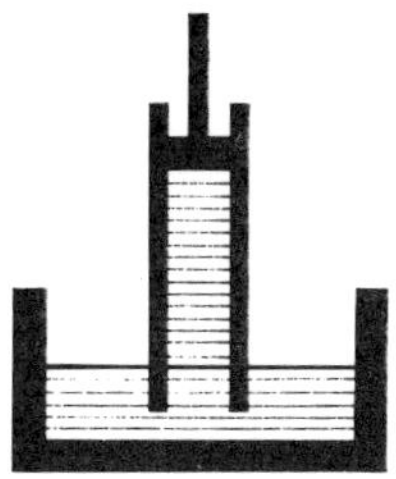

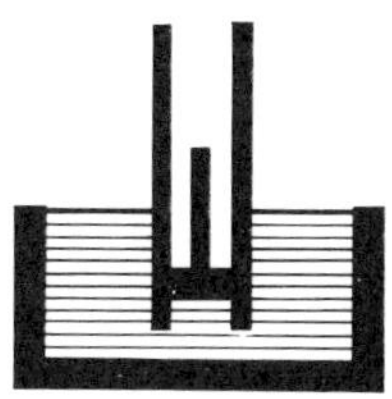

Try a simple flapper valve made from a flexible material. Water will now be trapped in the cylinder as it cannot return to the bowl.

What has to be added to transfer the cylinder water into a new container?

Look up the details of a living pump - the heart. What sort of valve does it have?

What qualities should a flap valve have?

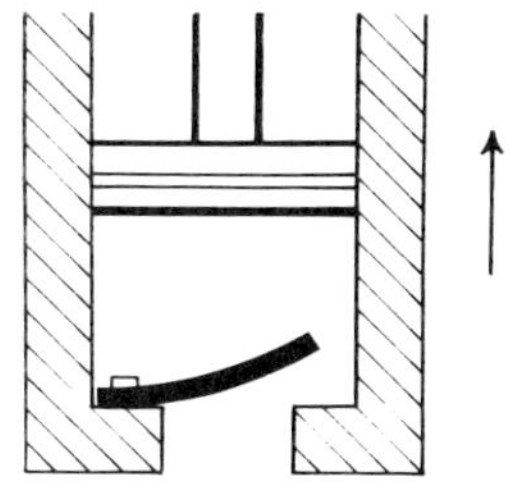

Valves can be made out of ball bearings. Football adaptors work in this way.

Ordinary steel ball-bearings are not as good as stainless steel ones - why?

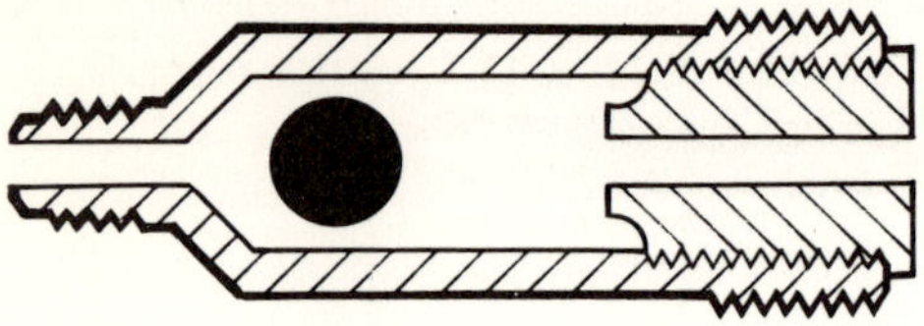

CONSTRUCTION

Suppose you wish to make a pump for a small garden fountain. The power source could be the wind. Refer to the chapter on Wind Power.

A small piston and cylinder pump might be suitable. The turbine drives a crank which pushes the piston up and down. The lower pipe is immersed in the water.

Write out your own design requirements. Then after discussion and reconsideration modify them and add more detail.

In the pump shown there are several serious faults. They include:
(a) the main crankshaft bearing is too weak;
(b) the windmill cannot turn into wind and if able to do so it would twist the pipes;
(c) the steel parts will rust.

Overcome these faults and design a better machine.

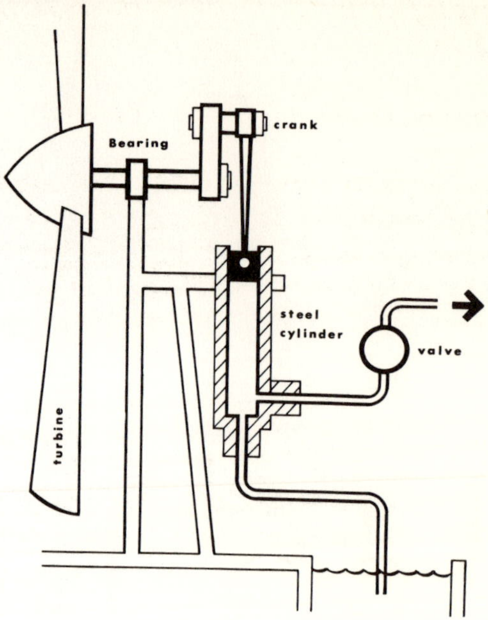

OTHER PUMPING DEVICES USEFUL IN THE HOME

Design and make a fish tank air pump.

Requirements: complete safety - no danger of electrocution. Complete reliability and little maintenance. No oil able to enter fish tank from the pump. No part of pump to be affected by corrosion, cheap to run, quiet.

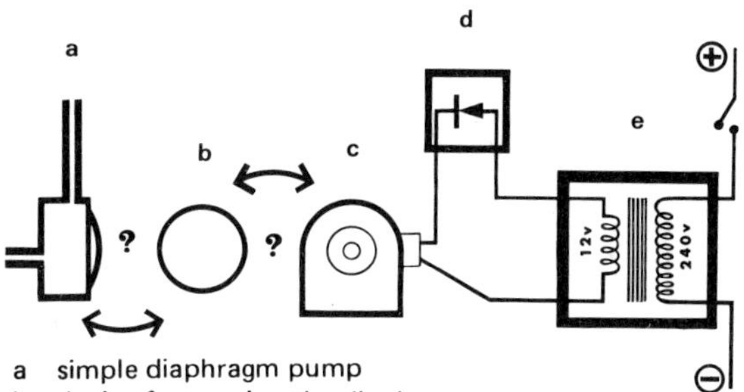

a simple diaphragm pump
b device for moving the diaphragm
c low voltage D.C. electric motor
d dry metal rectifier
e model train type transformer

Design and make a dispenser for washing-up liquid

Where will it stand? What size of container is it to fit? Is the amount of liquid dispensed to be adjustable? Is the liquid corrosive? What will power the dispenser, hand pushing, hand gripping, foot pedal, electricity, compressed gas (CO_2)?

Before you start, write your own brief putting the most important items first.

For example:

1. complete safety (wet hands will touch it);
2. simplicity and ease of operation;
3. high-pressure low-volume pump.

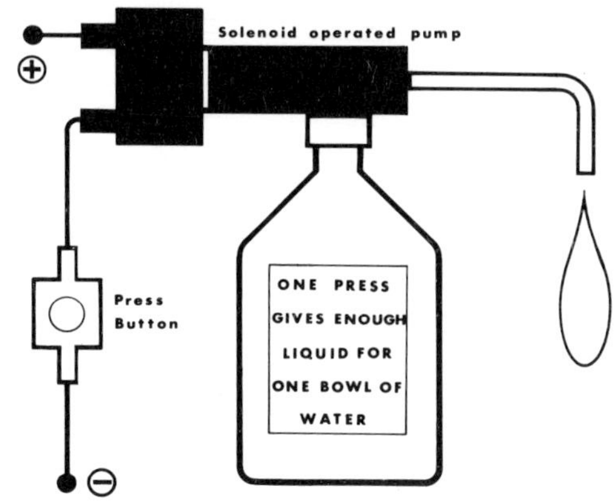

DESIGN GUIDE

Design and make an air pump for pumping up an air bed or an inflatable boat — low pressure — large volume. Here are four solutions.

a

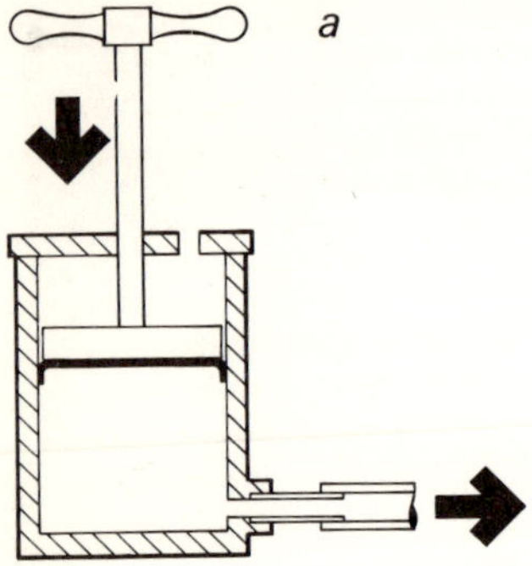

c

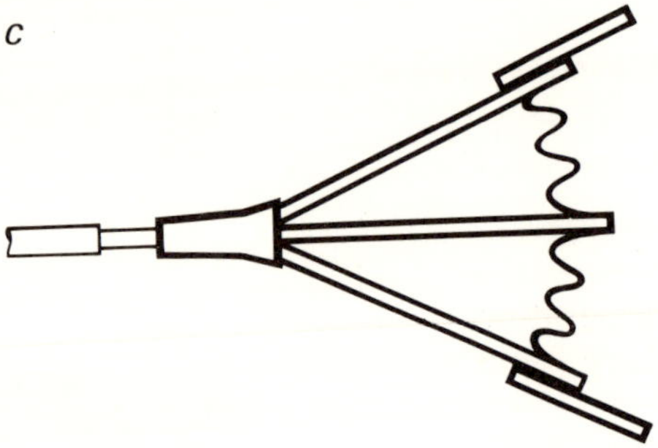

b

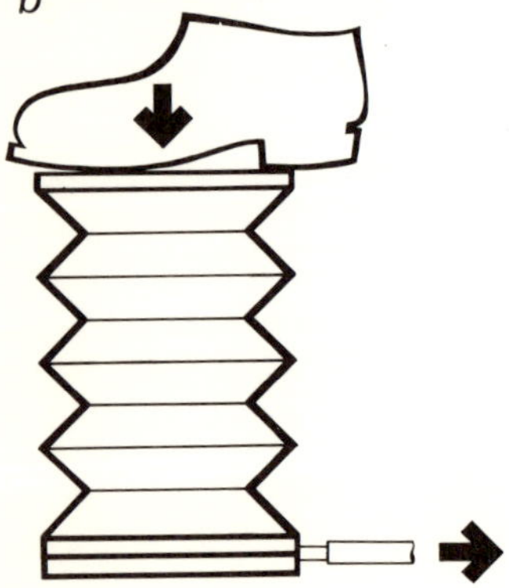

d

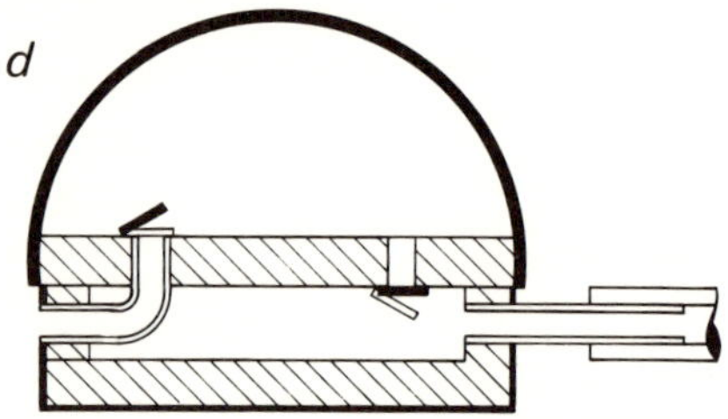

EXAMPLE

Suppose (d) is the most suitable for your requirements, that is, it fits the brief you have written most closely.

It could be made with the following materials.

Diaphragm — a rubber hemisphere (half a plastic or rubber football).
Base Box — plywood, glass reinforced plastic cast aluminium or sheet steel.
Valves — ball valves or flat plastic, rubber or leather flaps.
Hoses — rubber tubing.
Pipes — copper tubing.

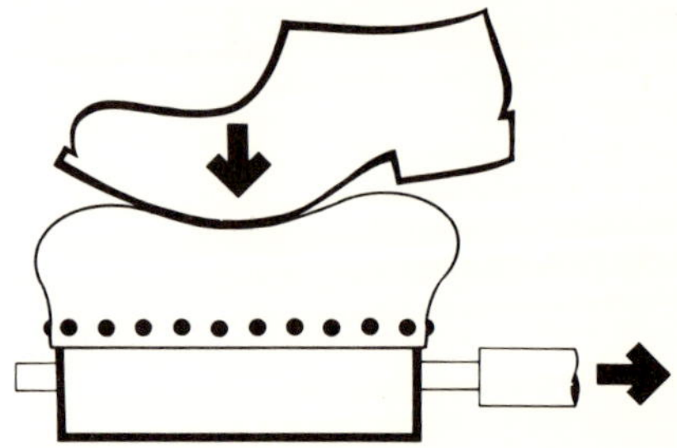

6 Wind power

Your aim in this section is to make the power of the wind work for you to drive a model, a dynamo or a pump.

This topic can be undertaken by a team of four or six pupils. It is important however for each member of the team to study the whole chapter so that a decision can be made on the following points.

(a) How simple or complicated a machine is required and what is to be its future.
(b) What help from specialist teachers is required.
(c) How the investigation and constructional work is to be shared out.
(d) Who is in charge of the project.

The earlier chapter on measuring the wind will provide useful information.

The first mention of a windmill, of the western type, is to be found in a Norman document of about 1180. Windmills were developed into important sources of power for grinding corn, sawing wood and pumping water, together with use in draining the land, working hoists in mills and mines, working ventilation devices.

A Dutch windmill built in 1783. *Photo by courtesy of Paul Popper Ltd.*

With present-day knowledge and materials it is possible to make quite small and efficient wind powered machines for generating electricity for a shed or cottage, or pumping water from one level to another, or for working a garden fountain.

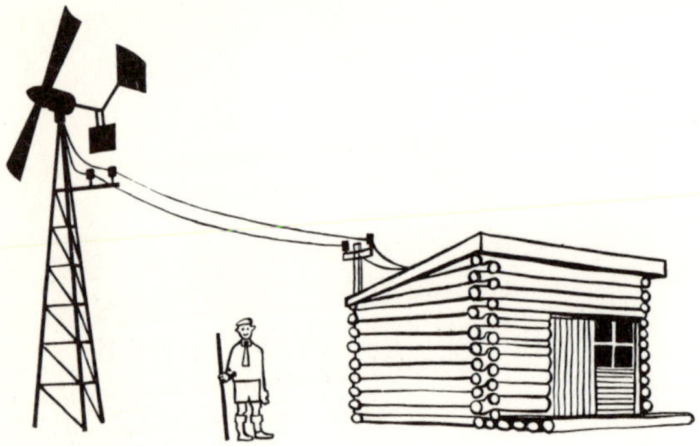

DESIGN GUIDE – MAKING A WIND POWERED MACHINE

Your aim is to make a wind powered dynamo fulfilling the following requirements:

1. to drive the dynamo at a suitable speed in average wind conditions;
2. cheap to make;
3. to be reliable and at least semi-automatic.

It must therefore:

(a) be completely safe;
(b) turn itself into wind;
(c) resist damage through gale force winds and exposure to poor conditions.

The more efficient the machine, the better it will be in terms of keeping down size, maintenance and cost.

It is assumed that you have a suitable dynamo (or pump).

HOW TO START

Two immediate problems face you – those of finding the speed and force required to turn the dynamo. The speed can be found in various ways:

Special test rig with watch and counter.

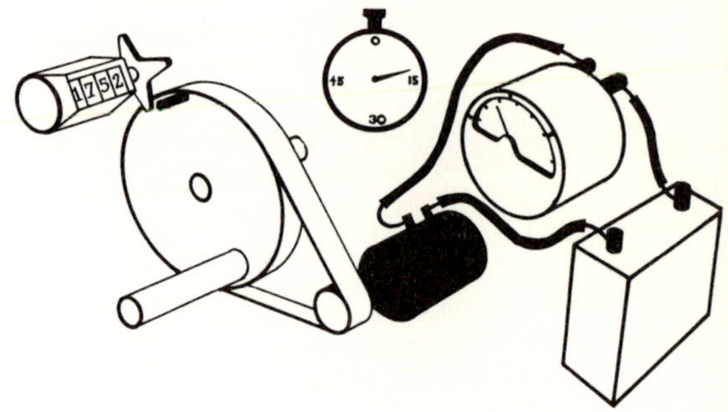

Bicycle on stand and stopwatch.

Lathe running at known speed.

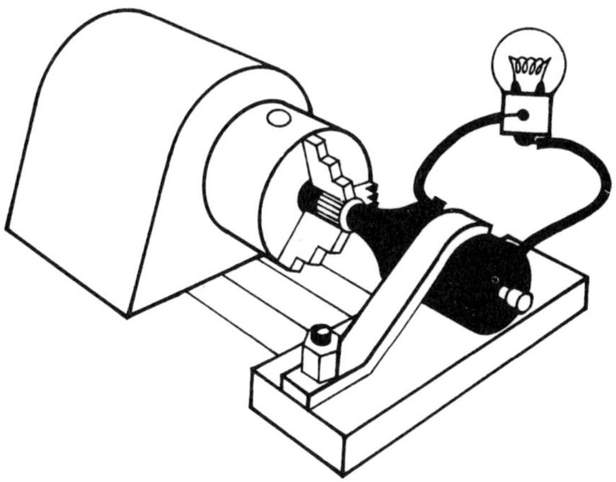

Force can be measured on a simple dynamometer. The reading on the balance (graduated in newtons) will give the turning force (when the dynamo is working) if multiplied by the length of the arm (in metres).

The balance weight allows the spring balance to read zero when the machine is stationary.

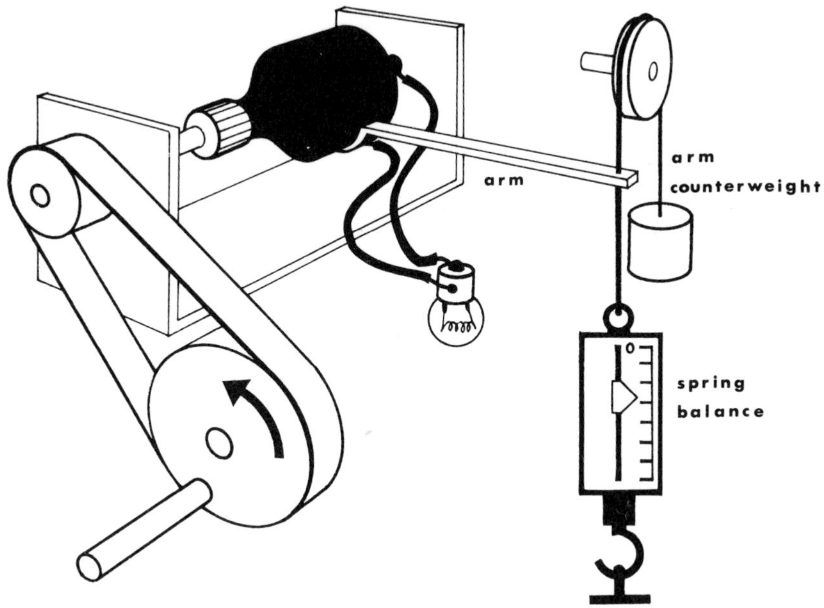

The next step is to find the average wind speed in your locality (see Chapter 1).

This is the speed at which your machine must be designed to run. The blades must be the right size to generate enough force to turn the dynamo.

The force generated depends on three factors.

1. The angle at which the blades are set (see page 31, blade angles).
2. The size and number of blades. Larger blades or more blades produce more power (see page 31, size and number of blades).
3. The aerofoil section used (see page 31).

HOW THE BLADES WORK

The blades can be thought of as a screw which travels forward as it rotates. Although windmill blades do not actually move forward, the effect is similar. The air passes through the blades providing the force to turn them.

A point on a blade will appear to travel in a screw path (or helix) through the air. θ is the pitch or helix angle.

This is $\dfrac{\text{distance travelled}}{\text{number of turns}}$

if all 'blade slip' is ignored.

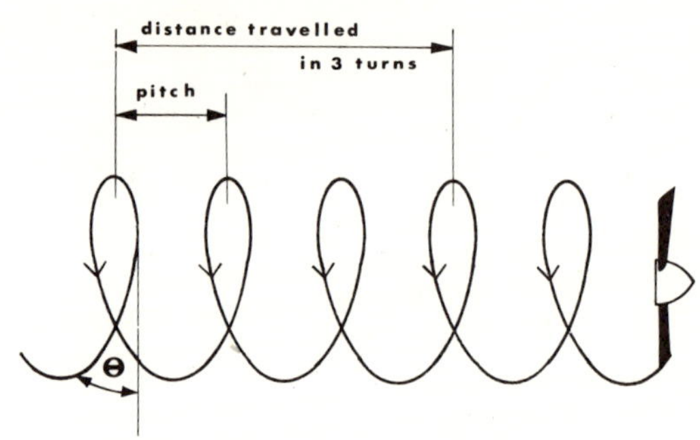

INVESTIGATING THE BLADE ANGLES

Do some trials to find the effect of wind on a steep blade angle (coarse pitch)

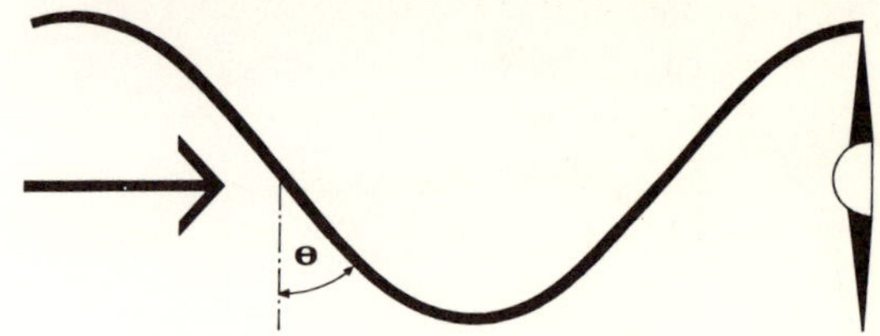

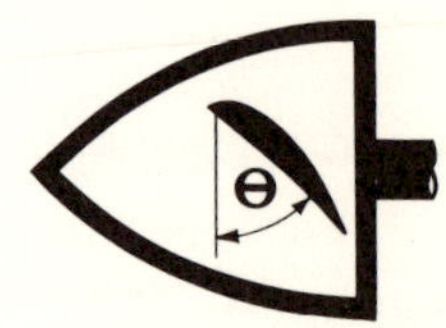

and on a shallow blade angle (fine pitch).

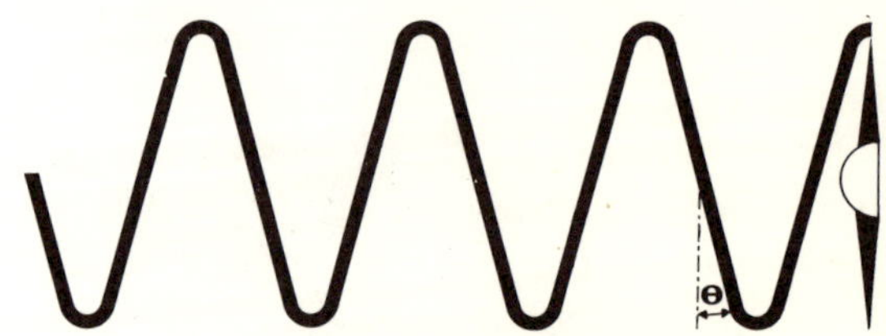

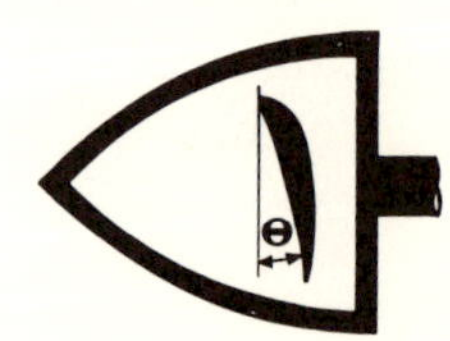

You will find that the dynamo needs to revolve at a greater speed than is safe for wind driven blades. Look at *Mechanisms* for suggestions on stepping up speed.

Make a decision on the speed you feel is best for your blades.

Check the centrifugal force which acts on a revolving blade; get your mathematics teacher to help you. (See page 68 for blade size.)

FINDING THE BLADE ANGLES

Suppose you decide on a blade speed of 300 rpm or 5 revolutions per second. This is a reasonable speed for windmill blades of between 300 and 800 millimetres radius.

You will have noticed that aeroplane propeller blades appear to be twisted, the outside of the blade being set at a much shallower angle than the inside. This is because the pitch (distance travelled forward per turn) is the same for the outside and inside of the blade, despite the fact that the outside of the blade travels round a much greater distance than the inside during one turn.

Look at these two cars climbing a hairpin bend. The car on the inside has to climb a steeper slope but at a slower speed than the car on the outside which has a gentle slope but must go faster to keep up with the inside car.

It is necessary to find out the angles to which each part of a windmill blade must be set. At the very least you will require the root angle and the tip angle of each blade.

You have chosen a blade which looks like this one.

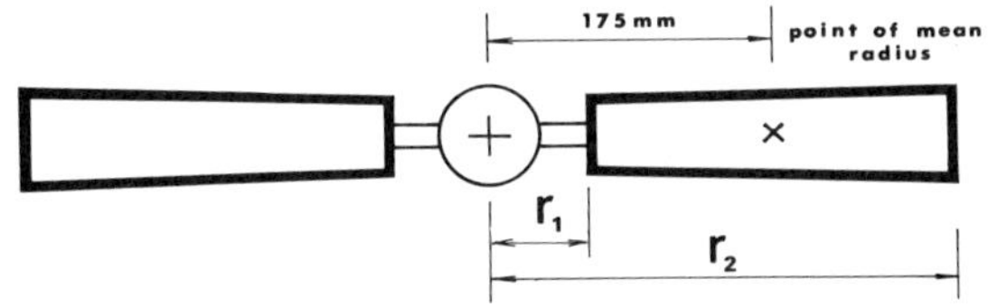

Here are the details.

r_1 is the root radius = 50 mm or 0.05 metres.

r_2 is the tip radius = 300 mm or 0.3 metres.

x is the average or mean radius and, for centrifugal force calculation, this is taken to be the position through which the mass of the blade acts.

Suppose the average wind speed is taken to be 10 metres a second and the blades are to revolve 5 times a second (300 rpm). The pitch of the blade will be the distance of air passing the blade in one turn = $\frac{10}{5}$ = 2 metres.

The tip of the blade will appear to travel in a screw path or helix in relation to the wind. You can draw out the helix on a cardboard tube and then peel it off.
The diameter $\times \pi$ gives the circumference. For r_2 this is $2 \times 0.3 \times \frac{22}{7} = 1.88$ metres.
A scale drawing will allow θ to be measured. In this example the tip angle (θ) will be 47°. (See pages 89–90.)

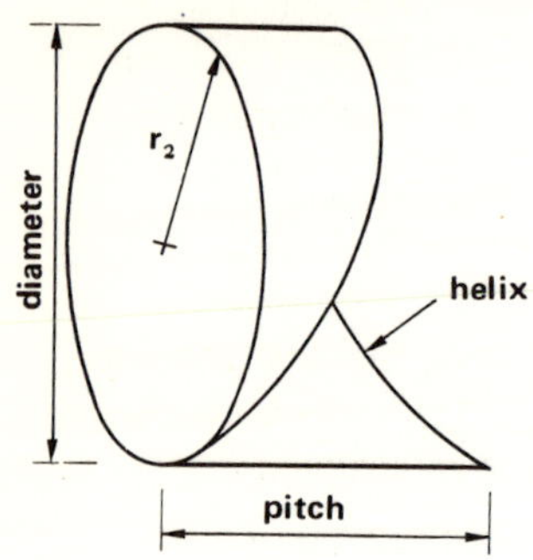

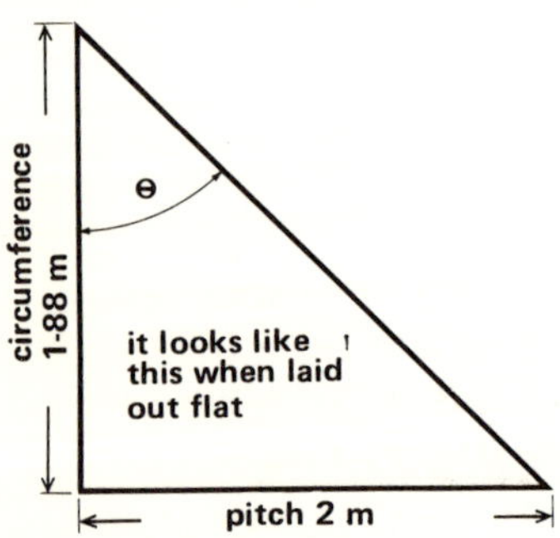

CONSTRUCTION OF BLADES

In the same way, by using scale drawings of the helix triangles, the angles for any other parts of the blade can be found. Look in the Appendix. A jig for mounting the blade ribs on is then made, each of the calculated angles being cut out of plywood and mounted to the correct spacing along a base board.

The changes in angle will be found to increase in almost equal amounts so the blade will be satisfactory if root and tip are calculated and the rest found by placing a straight edge along x to y. If the blade section is slightly modified to take round leading and trailing edges, then a built-up blade can be made using dowel rods for the edges thus removing the need to twist specially-shaped members into position. This prevents the building-in of stresses into the blades. A modified R.A.F. 31 areo-foil section is very suitable. Having solved the blade angle problem the next step is to build some trial sets of blades of spans say between 300 and 500 mm. The chord will have to be estimated and, during trials, should be kept the same for all the blades. A figure of about 75 mm is suitable for trial blades. As the torque of the blades is dependent on surface area and diameter, a decision has to be made about the number of blades used. Find out whether to use 1, 2, 3, 4, 5, or more blades. When considering the problem remember that:
(a) a large span gives greater power, is subject to large centrifugal and gyroscopic forces, sweeps a large area, and is harder and more expensive to make than a small blade; (b) a large number of blades provides more area for the wind to strike; each blade has to be carefully matched in shape and weight and has to be mounted in a hub sufficiently strong to carry all the blades satisfactorily; (c) odd numbers of blades are not always easy to mount with accuracy.

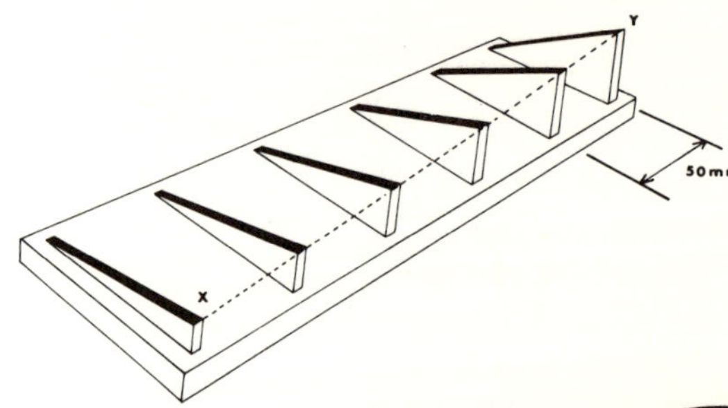

INVESTIGATING BLADE POWER

Test the trial blades for power output.

Each set of blades should be mounted on a test rig and their output measured on a rope dynamometer.

The reason for not allowing the blades to drive the dynamo directly is that the first trials with the blades will give results outside those possible to record on the dynamo. In other words it will be either too weak or too powerful and only after several trials will it be possible to interpret the results and modify the most suitable set of blades to give the required output at the correct speed.

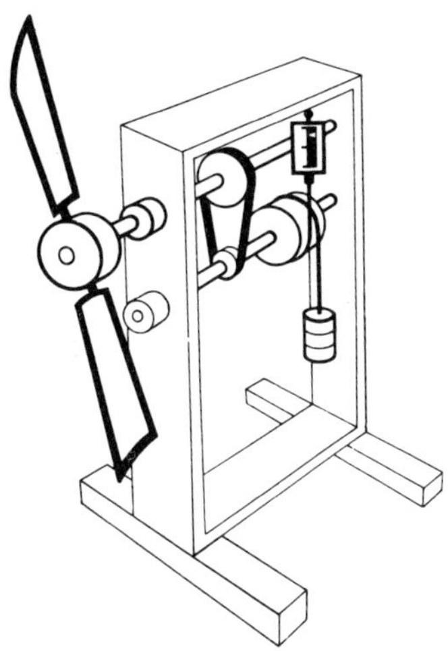

A means of moving air past the blades at a steady known speed will be required. It is unlikely that a wind tunnel will be available but other ways can be found. In 1795 John Smeaton conducted some experiments on windmill blades using a whirling table. You could follow his example.

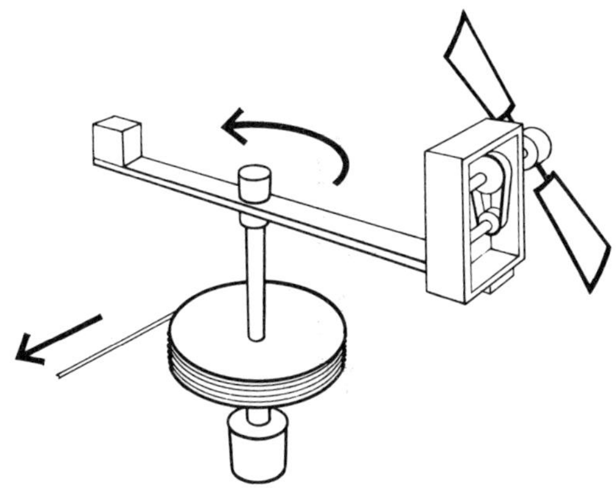

Another way would be to set up the rig on a vehicle such as an open car. In still conditions this would give a constant air stream of known velocity past the blades. An anemometer will be very useful for finding the wind speed when still air conditions are not prevalent.

For the blades to be really efficient a suitable aerofoil section should be used. As mentioned earlier, one suggestion is to use R.A.F. 31, which is a flat-based section easy to construct. Other suitable sections can be found in textbooks on aerodynamics.

The blades have to be carefully matched as stated earlier. For this reason, and for lightness, hollow shapes are more suitable than carved solid ones for large blades. Experiments should be tried using balsa-skinned blades, or blades made of polystyrene and other plastics such as glass-reinforced resin.

SAFETY

It is advisable to incorporate a steel spar with a strengthened root in built-up hollow blades. The centrifugal force resulting from the turning of the blades can be considerable and a weak hub or blade could easily break up when revolving at high speed. It is necessary to incorporate some governing device to stop the blades over-speeding in a gale.

Here are some ideas which might help you to solve this problem.
(a) Windmills used to be 'furled' like sails (the blades were called sails).
(b) It is possible to turn the blades out of the wind direction to regulate their speed.
(c) Aeroplane propellers can be twisted (or feathered) so as to match the conditions under which they are working.
(d) Aerofoils can be 'stalled' with air brakes.
(e) A centrifugal governor can provide the means of actuating the feathering or braking device.

Make sure that your regulating device will 'fail safe' as a runaway windmill is dangerous and will almost certainly break up. Remember to make provision for any change in wind direction and, in thinking about this, devise a method of preventing the electrical wires becoming twisted between the machine and its mounting.

EXAMPLE OF A WIND POWERED GENERATOR

This machine fulfilled its design brief. The requirements were:

(a) to generate 150 watts in a wind speed of 7 metres/second;
(b) to be made using normal school equipment;
(c) to be self-regulating to avoid over-speeding;
(d) to be self-aligning into wind;
(e) the entire machine to cost less than half the price of a new car generator.

The investigations included:

1. air flow over blades;
2. dynamometer tests on generators — input/output;
3. wind conditions;
4. methods of gearing (wheels, pulleys, friction drives);
5. methods of speed control (linkages, cams, governors, air and mechanical brakes);
6. methods of fabrication, machining and casting;
7. methods of electrical transfer, types of brush and slip ring arrangements.

The blades were made from plywood ribs on a steel main spar covered with balsa sheet. The main body is a two-part aluminium casting. The bearings, generator, gears and steel for main shaft were obtained from scrap yards.

A wind-powered generator made by pupils of Handsworth Technical School, Birmingham.

7 The theodolite

Study the next three sections and design and make an optical instrument of your own choice.

Read the whole chapter, make some trial drawings to try out methods of measuring angles and distances. Check your solutions using trigonometry. Discuss the meaning of elevation, bearing, level, horizontal, vertical, plumb, true, perpendicular, sighting, focusing.

Section 1

HEIGHT MEASUREMENT

It is not always possible to make measurements with a rule or even with a long measuring tape. For instance, if we need to find the height of a church spire it would be impractical to measure it directly.

If the top of the spire is sighted from a position which is a known distance (d) from the middle of the church tower, the angle of elevation will give the additional information required to allow the height to be found by making an appropriate scale drawing.

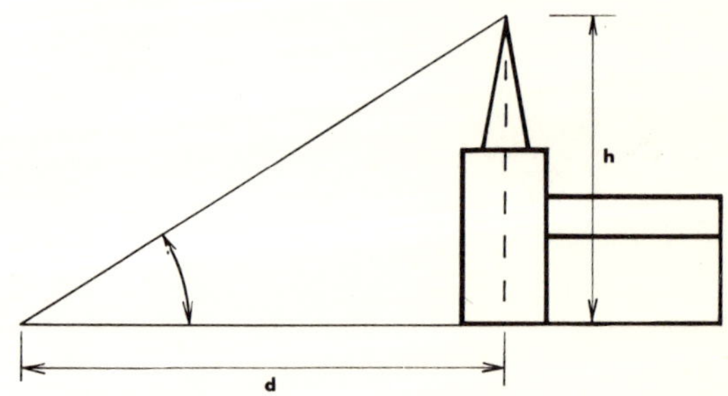

Try this idea out by making scale drawings of situations you can check by measuring. See how a scale right-angled triangle can be drawn using the following information:

(a) its height and base only;

(b) its base and base angles only;

(c) its height and base angles only.

Similar problems in measuring angles are encountered by navigators who use the angle of elevation of a known star, or of the sun, to calculate their position at a particular time. The instrument they use for 'shooting the sun' is called a sextant.

INVESTIGATION

Here are some of the problems which it is necessary to solve when designing a simple angle measuring instrument (a theodolite). We will assume for simplicity that the ground is horizontal.

The device has to be mounted on a base which is steady and horizontal. Ask yourself: how high should it be, can the height be adjusted? If supported on legs how many should be used? How can the legs be fixed to the base, and should there be a hinging device which allows the legs to be opened to a predetermined angle? How can the accuracy of the horizontal base mounting be checked?

The method of sighting must be accurate. A 1° mistake in looking at an object 100 metres away will result in an error of 1.75 metres. Is this an acceptable error?

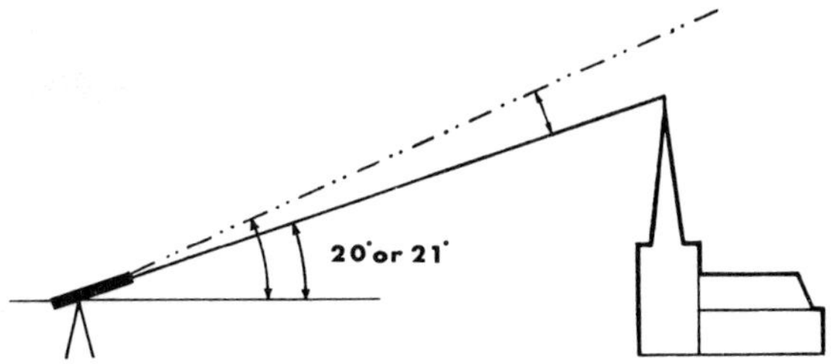

Just looking through a tube is not good enough — try it and you will see something like this.

There are ways of improving on this method of sighting.

The sighting device has to be adjustable in order to be capable of measuring different angles of elevation. A useful method is to pivot the sighting device on trunnions in the same way as ancient cannons were mounted.

The questions you must answer are: how can the sight be easily moved and yet remain exactly in the position required when aligned on the object being viewed? How can the angle of elevation be measured?

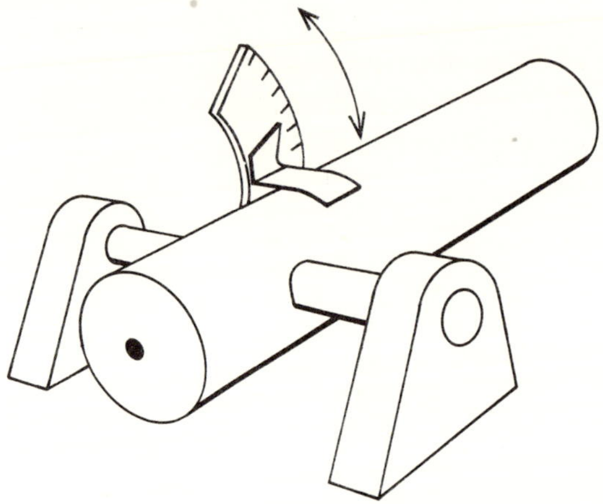

CONSTRUCTION

A mock-up in cardboard and wood can be made to test out your ideas. A permanent instrument will have to be made from materials such as metal and plastics in sheet, round and tube form. You will have to think about the methods you can use to shape, join and protect the parts, about the cost and about the time and equipment necessary for carrying out your construction.

In the theodolite shown opposite the sighting tube was made from metal tubing with perspex 'lenses'. All the other parts were made from mild steel rod or sheet.

Theodolite made by pupils of Handsworth Technical School, Birmingham.

Section 2

SURVEYING BY TAKING BEARINGS

Suppose you wish to do some surveying by 'triangulation'. Here is a five-sided field. Two points A and B are marked some convenient distance apart – say 50 metres. A theodolite is used to measure the angles between A and B and each corner of the field in turn.

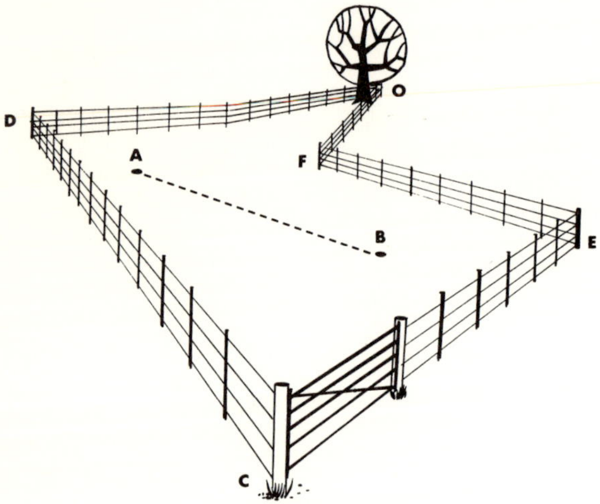

Take the corner where the tree is growing, for example.

The angle BAO is measured from A, and the angle ABO is measured from B.

This action is repeated for all the corners of the field and a scale drawing (map) of the area is drawn by joining up OD, DC, CE, EF and FO.

The following angles have to be found: BAO, BAF, BAE, BAC, ABO, ABF, ABE, ABC.

The distance DA which is in line with AB has to be measured.

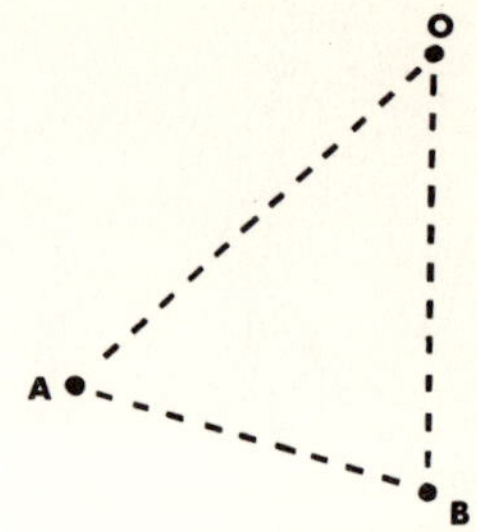

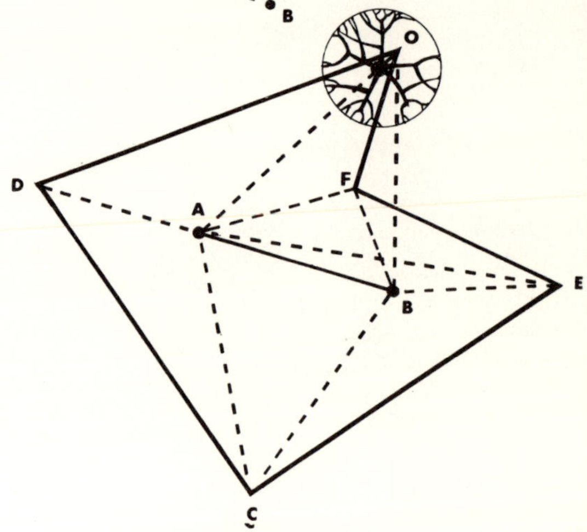

INVESTIGATION

You will have to modify your theodolite in order to make these measurements of bearings. What modifications are needed, how will you carry them out? Is it possible to make a survey on sloping ground? Consider how you overcome the problem of not having a horizontal line to use for the base of your triangle.

CONSTRUCTION

When you have solved the problems connected with the principles, selection of materials and shapes (the drawings show some of the shapes of parts which you might have chosen), then the constructional methods will have to be considered.

If the parts are made from sheet metal you will have to:

1. Shape them accurately.
2. Mark out and drill the holes accurately.
3. Fold where necessary, at the correct places.

If the parts are made from tube, rod and flat sheet it will be necessary to do the following.

4. Face off the ends of the tube and rod.
5. Align and fasten the trunnions to the tube without leaving the trunnions across the centre of the tube.
6. Make the protractor so that when fixed to the tube it is accurately positioned.
   The centre point of the semicircle should intersect the centre line linking the trunnions but no part of the protractor should be inside the tube.
7. Make a suitable sighting system for the tube.
8. Have a device to hold the tube in any desired position.
9. Fix scales and pointers for reading vertical and horizontal angles to the protractor and base plate.
10. Fix the base plate to suitable legs.
11. Protect the instrument from corrosion.

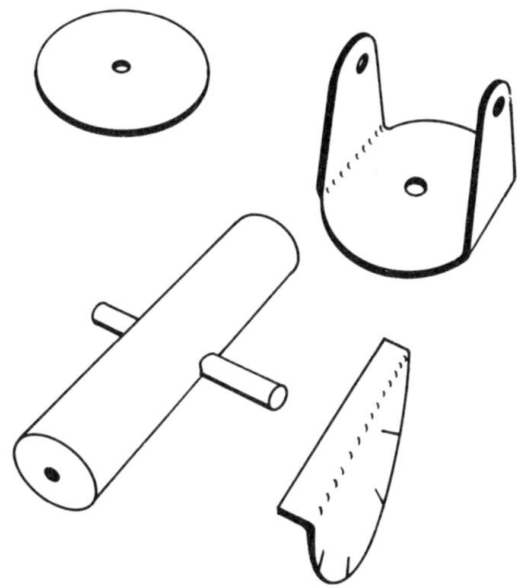

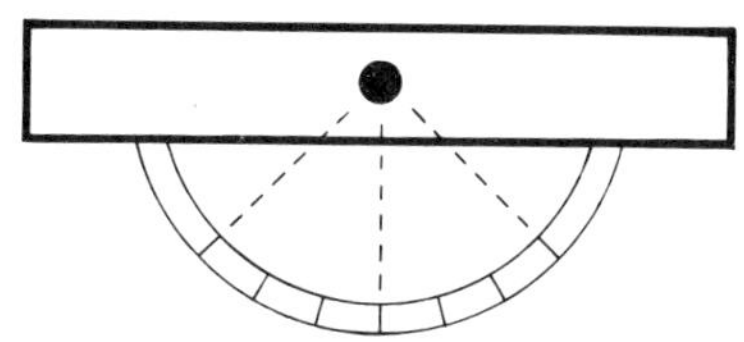

8 The periscope

A simple periscope is easy to make and to understand. A more complicated instrument is much more difficult to understand and to make, and is much more costly because it incorporates optical devices such as lenses and prisms instead of mirrors. The periscope of a submarine is an advanced piece of optical engineering resulting in a robust, large but very precise instrument.

INVESTIGATION

Here are some experiments to try. You will require two reasonably good small mirrors (ladies' handbag mirrors will do), a flat board and some blocks of wood for holding the mirrors.

First set up one mirror and make some observations.

If the mirror is upright you will be able to see an object (x) by looking in the mirror. The angles R and I will be the same size, but the image will be reversed.

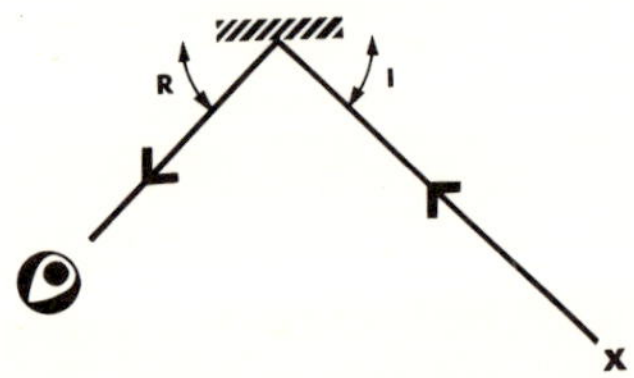

Have you ever seen this word printed on a vehicle? Why is it like this?

Place a second mirror so that it picks up the reflection from the first mirror.

Try varying (a) the distance between the mirrors, (b) the angle of the object to the first mirror, (c) the angles between the mirrors. What effect have different arrangements on the field of vision (how much can be seen)? What is the effect of using two convex (car rear view) mirrors in your periscope?

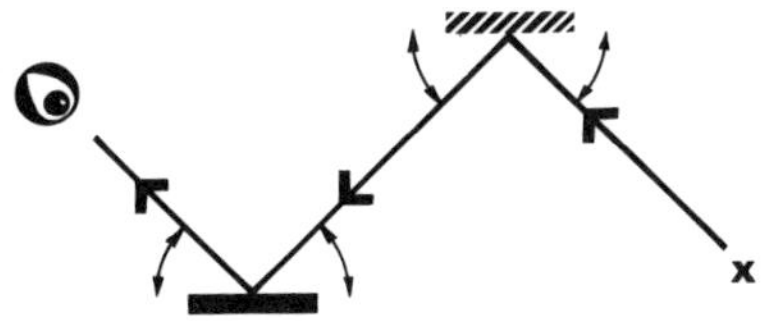

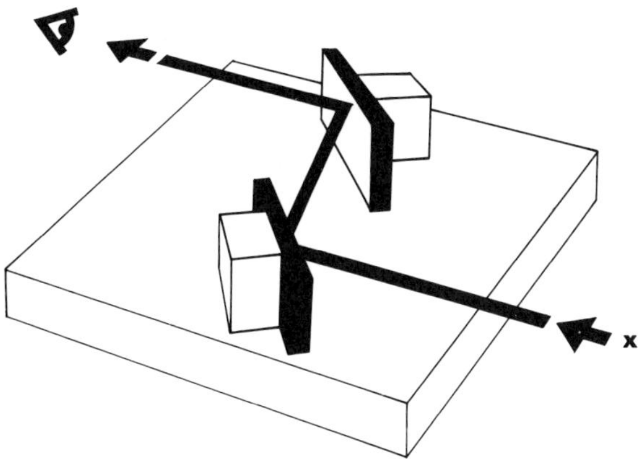

You can try different sorts of mirror, made from glass, polished stainless steel, coated plastic and chromium plate. Probably the best mirror, apart from a prism, is a glass plate coated on the front with aluminium or silver. This type of surface is used on high quality instruments such as reflecting telescopes and on camera viewfinders.

After the first tests a cardboard tube will enable you to make a mock-up of an actual instrument.

The following problems will need attention: the cross-section shape of the tube; the internal colour and surface – shiny, dull, white, black; should the mirror distances or angles be adjustable; what is the best shape for the apertures; how is the instrument to be supported – a stand, handgrips, suspended from an overhead member?

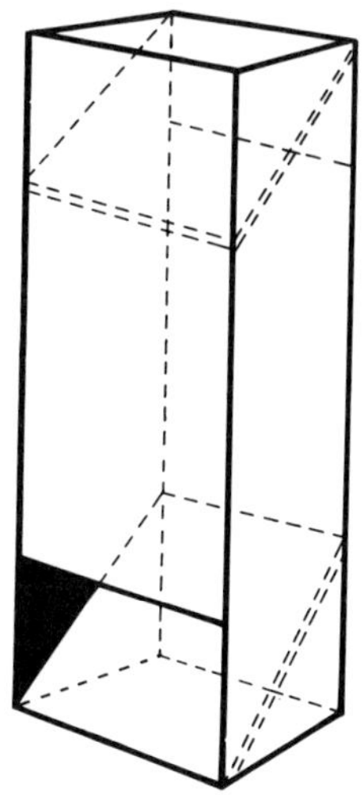

If the periscope is to be used for, or as part of, a direction-finding device then a sighting device will have to be fitted.

A sighting telescope with a pinhole at the eye end and a cross at the other is suitable.

In a sighting telescope extremely fine lines are placed between the lenses at the point where the real image is produced by the objective lens. Such an instrument is employed on commercially-made theodolite but a simple instrument without lenses will work well if made carefully. Here are some further possibilities for sighting systems.

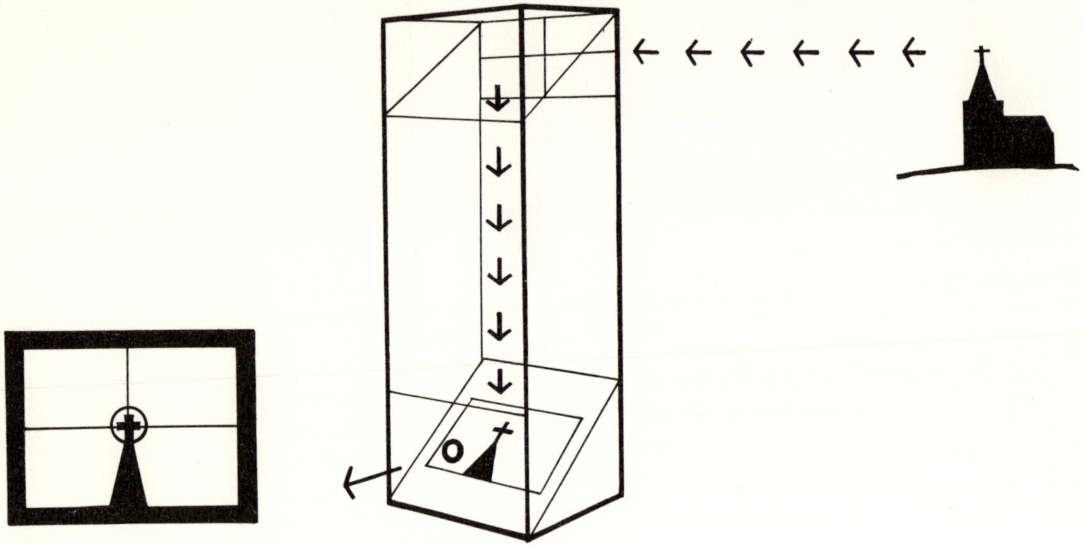

fine cross lines on mirrors

circular windows

pin heads in tube

ring and pin

CONSTRUCTION

A good periscope can be made from soft aluminium sheet of a thin gauge, the joints and seams of the case being bolted, pop riveted or screwed with self-tapping screws. The tube shape can be formed over an accurately-made wooden block. Pushing into shape, rather than hammering, is recommended to avoid hardening the material so that it becomes springy. Hammering also results in unsightly dents along the folds in the material which spoil the appearance of the periscope. A 'hammer finish' paint will give the case an acceptable appearance.

9 The range-finder

When you have experimented with the periscope and the theodolite you might like to use the principles employed in both of them on one instrument. This device could be used to find very quickly how far an object is from your eye. Such devices are used with cameras and guns to measure the range of objects viewed through the sight, thus enabling sharp focusing or accurate shooting to take place. Triangulation is again the principle being used. If you know the length of one side of a triangle and the size of two of its angles then the length of the other sides and the third angle can be found.

In the measurement of height AB is known, angle CAB is known, angle ABC is 90°. So CB (and AC if required) can be found by drawing the triangle.

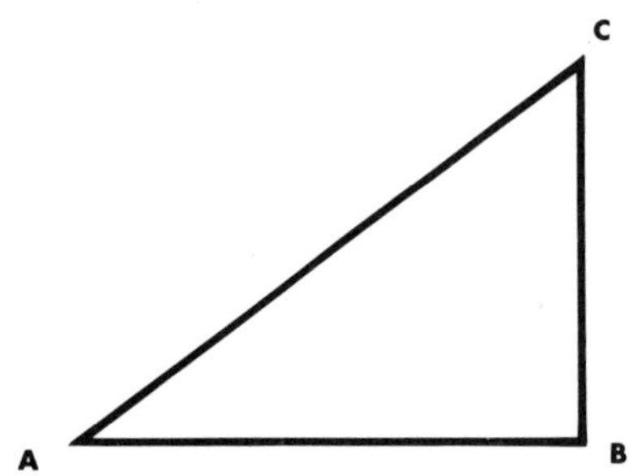

In the survey by triangulation AB, and angles BAO and ABO are known; by drawing the triangle to scale all the other distances can be found.

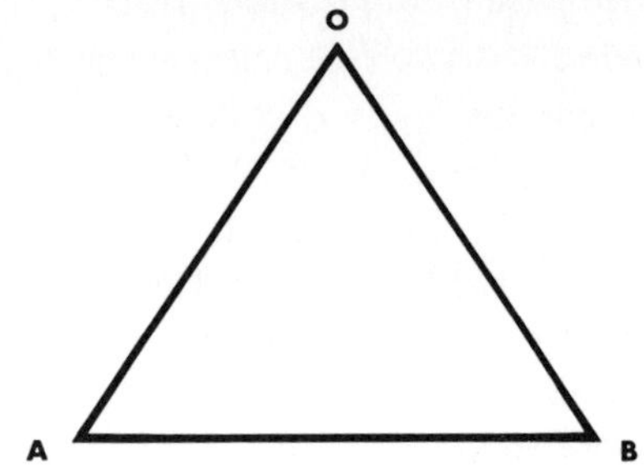

Imagine you are looking down on a gun emplacement from an observation aircraft.

The gunner has to know the range, as well as the direction, in order to hit the target. Too many trial shots would reveal the position of the gun.

A good method of finding the range would be to take sightings from two positions.

If the distance from the gun W and the sighting position Z were measured before the battle and the angles WZT and ZWT were measured during the battle then the distance WT could be easily found. If the target T was fixed then angle ZWT could be made 90°.

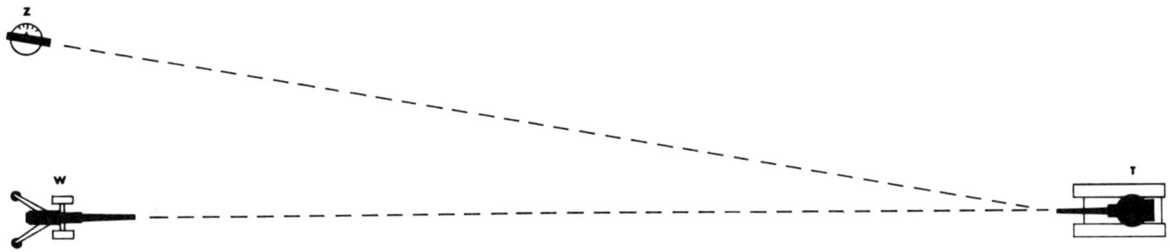

To avoid large adjustments in the gun's bearing, especially when tracing a moving target, two sighting positions spaced at each side of the gun would be used.

BW and AW would be known. Angles BAT and ABT would be measured. A scale triangle of ABT could be drawn and distance WT found.

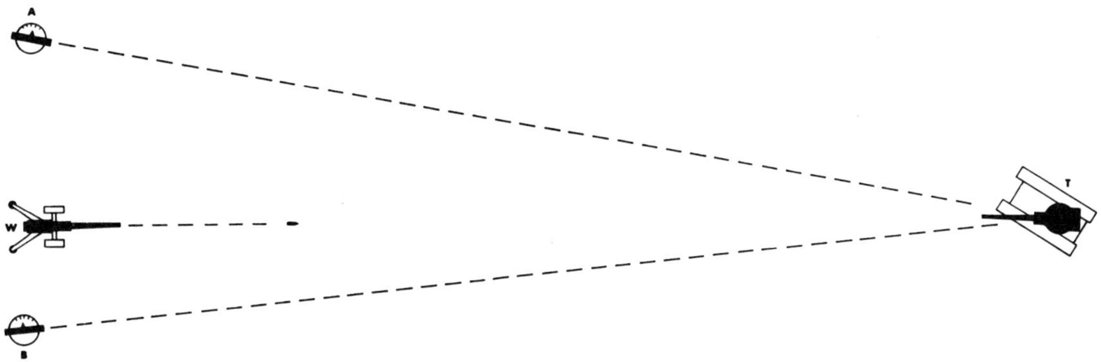

INVESTIGATION

The smaller the distance AW and BW the more accurate the angle measurements must be. Can you find a way of working out the range using this principle, but avoiding the need for men at A and B? If there was only one man, placed at W, what could you place at W, A and B to measure angles BAT and ABT?

CONSTRUCTION

It is possible to buy very compact and accurate rangefinders suitable for use in photography. These are usually expensive as they employ lenses, prisms and mirrors. If you have understood how the ranges are found in the examples given in the previous pages, you will be able to design a simple rangefinder which, although not as compact as a professionally-made instrument, will work very satisfactorily over a moderate range and will cost very little.

Here is one solution; you could make a mock-up first to try it out.

AB is known (1 metre?). Angle ABT is 90°. Angle BAT can be measured.

Could the ranges be read directly off the protractor at A?

By turning the whole instrument and sighting through B, angle ABT must become 90°, angle BAT is then found, and the distance of T established by scale drawing.

The parts are easily made from wood, plastic and metal. The methods for measuring angles, obtaining accurate sights, locking the moving parts, and folding up for carrying, are all similar to ones employed on the periscope and theodolite.

If a mirror system is employed, sights A and B could be coupled to a central control and both moved into position simultaneously. This would enable the base AB to be fixed down. The methods described are only simple ones; modern gunnery sighting is done by radar and can allow for gun movement, target movement and other factors such as wind direction and force. Your instrument could be used for measuring shot and javelin throws, if you make distance AB sufficiently large for purpose of accuracy.

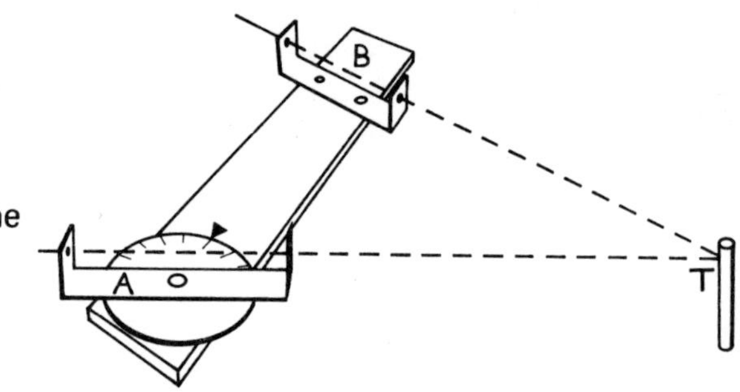

Materials

When trying to decide which materials to use to make an object you must remember that your choice will be affected by several factors such as availability, ease of working and cost, as well as by the special factors of strength, appearance, wear and corrosion resistance, etc. In order to help you to decide which materials and shapes to use, the following notes are given.

The materials mentioned are ones which might be found in most school craft rooms and home workshops. The notes are intended as a guide to selecting a suitable material. They are not complete accounts of the properties of materials or of the correct methods of working them. Further information should be found from other sources if needed.

1. FOR CONTAINERS AND CASES

Sheet materials such as tinplate, steel, brass, copper, zinc, aluminium, acrylic (perspex), plastic laminates, ebonite, tufnol, paxolin, plywood, hardboard.

Hot formed shapes

Acrylic sheet, polystyrene sheet, to be blown or pressed into shape.

Cast shapes

Zinc alloys, resins reinforced with cloth or glass fibre, papier mâché.

2. FOR LIGHT FRAMEWORKS

Mild steel or aluminium alloy tubes, strips, and other sections such as angle, tee sections, channel rolled or extruded into shape. Wood such as ash, bamboo, or spruce.

3. HEAVIER FRAMEWORKS

Steel joists, angles, channels or heavy gauge tubes (scaffold poles). Thick sections of timber, reinforced concrete.

4. TOOL EDGES

High carbon steel, piano wire, silver steel.

5. SPRINGS

Strips and wire in high carbon steel and phosphor bronze.

6. BASE PLATES, LINKS, CYLINDERS, PISTONS

Castings in zinc alloy, aluminium, brass and iron; strip, rounds and plates of mild steel and brass.

7. BEARINGS, PLAIN BUSH

Castings, tubes and rods of bronze, brass, carbon steel, aluminium, cast iron, nylon and blocks of lignum vitae.

8. AXLES, FORGED, TURNED OR GROUND TO SHAPE

Silver steel, high carbon steel, alloy steels, mild steel. Please note that any load-bearing axle, for instance a bicycle spindle or go-cart axle, should be made from special alloy steel made for that purpose and which is given accurate heat treatment before use. Axles made from other materials might easily break or bend under shock. High carbon steel and silver steel is only suitable for axles on models and instruments and should be hardened and tempered. Mild steel is suitable for toys and models if case hardened before use.

Read the section on friction in the chapter on structures before selecting an axle and bush.

9. CORROSION-RESISTING MATERIALS

Pure iron, pure aluminium, copper and its alloys, zinc and its alloys, lead, silver and its alloys, perspex, plastics, resins, glass and ceramic materials, ebonite, tufnol, paxolin and stainless steel.

10. MATERIALS WHICH ARE EASILY CORRODED BY THE ATMOSPHERE

Steel, duralumin.

11. ELECTRICAL INSULATORS

Ebonite, wood (lignum vitae, black bean), paxolin, tufnol, ceramics, perspex, plastics, rubber and glass.

12. ELECTRICAL CONDUCTORS

All metals mentioned previously, copper, silver and aluminium being especially good. Carbon also is a good conductor of electricity, water will conduct electricity if impure, dilute sulphuric acid is used as an electrolyte in accumulators.

13. HEAT-RETAINING MATERIALS

All the metals mentioned previously together with water, firebrick, brick, stone and concrete, copper being exceptionally good.

14. HEAT-CONDUCTING MATERIALS

All the common metals, the best being aluminium and copper. Most dense materials will conduct heat to some extent, for example, concrete, brick, glass. Water and other liquids, which should be non-inflammable and of a high vaporising temperature, will conduct heat if convection is allowed. Air will conduct heat by convection.

15. HEAT-INSULATING MATERIALS

Any material which is not damaged by the heat it is subjected to, and which is of a low density, will give good insulation. A material which traps air and thus prevents it circulating, is usually a good insulator. Wool, cork, expanded polystyrene and insulation board, are all good insulators. For temperatures approaching the melting points of metals and those of fire and flame, some asbestos materials or firebricks will be suitable.

16. SOFT SOLDERING MATERIALS

Iron, steel, copper, brass, bronze, gilding metal, tin, zinc with difficulty and aluminium only with special solder and flux.

17. HARD SOLDERING METALS AND ALLOYS

Silver, steel, copper, gilding metal, brass can be hard-soldered.

18. BRAZING METALS AND ALLOYS

Steel, copper, gilding metal can be brazed.

19. GLUING OR BONDING MATERIALS

(a) Epoxy resins, of which there are many different kinds, are usually expensive but will fasten wood, metal, glass, ceramics, leather and plastics with great strength.
(b) Contact adhesive, a rubber solution type of adhesive, is moderate in price and will fasten wood, plastic laminate, metals, glass, and cloth with reasonable strength.
(c) Resins for bonding glass fibre reinforcing materials together have great strength except under impact, are easily obtainable, and require mixing with accelerator before they are used.
(d) Resins for timber: usually water-proof and of great strength. They have a limited storage life, are easily obtainable and require mixing with water or with a hardener before use.
(e) Polyvinyl acetate for timber and paper: not water-proof but very strong and slightly flexible. They have a long storage life, are moderately priced, and are very easy to apply, no mixing being required.
(f) Cellulose-based paper glues, for example wallpaper paste: these are not usually water-proof but are cheap and easy to use on paper. They require mixing with water before use.

20. METHODS OF FASTENING

Bolting, riveting, wedging, strapping, welding, brazing, hard soldering, soft soldering, binding, jointing, gluing, bonding, pinning, nailing, screwing.

21. *METHODS OF FORMING AND THE MOST SUITABLE MATERIALS FOR THESE METHODS*

Bending - mild steel, wrought iron, annealed carbon steel, copper, lead, zinc, aluminium.
Casting — lead, aluminium, brass, bronze, cast iron, zinc alloys.
Pressing —mild steel, copper, aluminium, brass.
Extruding — aluminium, copper.
Drawing — steel, copper, silver.
Beating — copper, gilding metal, steel, brass, aluminium.
Forging — wrought iron, mild steel (copper, silver and aluminium can be cold forged).
Machining — brass, bronze, aluminium, mild steel, free cutting steel, annealed carbon steel, cast iron.
Cutting with hand tools — all mentioned previously except hardened carbon steel and case hardened mild steel. Materials like glass can be cut with tipped tools or ground to shape with abrasives.

22. *PROTECTIVE PROCESSES*

Metal plating with copper, zinc, tin, silver, nickel, chromium, coating with shellac, polyurethane varnishes or cellulose lacquer. Painting with oil or cellulose based paints, coating with grease or oil. Ship fittings are often protected by the method of 'sacrificial corrosion'. A block of a metal which is higher in the electrochemical series than the metal part of the ship is placed below the waterline of the vessel. This material is then corroded away, instead of the ship, by the electrolytic action of the two different metals and the sea water. From this example it can be seen to be unwise to place two different metals together in a situation of exposure to moisture. For example, a house water system containing copper and galvanised iron tanks will have serious corrosion problems.

23. *FATIGUE IN MATERIALS*

If a piece of wire is repeatedly bent backwards and forwards at the same place it will break. An examination of the fracture under a microscope would reveal a different crystal structure from that present at other parts of the wire. Some materials are more susceptible to breaking in this way than others. The cause of failure is often called 'fatigue' and is closely connected with the work-hardening property of a material. Highly stressed parts, such as airframes on aircraft, are more susceptible to fatigue than modestly stressed components such as the frame of a railway locomotive. Some materials are more likely to fatigue than others; when working material of this type, heat treatment is necessary to avoid difficulty and damage. Aluminium, copper, brass and steel should all be given treatment if subjected to constant bending, beating or other forceful forming operations.

Appendix

LINEAR EXPANSION OF MATERIALS

100 mm of the following materials will expand lengthways the amount shown for every 5°C rise in temperature.

Mild steel	0.006 mm
Copper	0.008
Brass	0.0095
Aluminium	0.012
Glass	0.0045
Wood	0.0015
Nickel steel	0.000045
Lead	0.015
Zinc	0.013
Stone	0.0025

EXAMPLE

A piece of aluminium 100 mm long at 25°C will become 100.012 mm at 30°C.
or
A strip of steel 100 mm long at 10°C will become 100 + (0.006 X 3) = 100.018 at 25° or
A strip of copper 200 mm long at 5°C will become 200 + (0.008 X 2) = 200.016 at 10°

TABLE OF ANGLES FOR WINDMILL BLADES

For blades to revolve at 300 rpm (or 5 revolutions a second).

Particular radius of blade in mm	Wind speed metres per second		
	5	10	15
25	81°	85.5°	87°
50	72.5°	81°	84°
75	65°	77°	81°
100	58°	72.5°	78°
125	52°	68.5°	74.5°
150	47°	65°	72.5°
175	42.5°	63°	70°
200	38.5°	58°	67°
250	32.5°	52°	62.5°
300	28°	47°	58°
350	24.5°	42.5°	54°
400	22°	38.5°	50°

EXAMPLE

For a windmill blade to revolve at 300 rpm in a wind speed of 5 metres a second with the blade tip at a radius of 300 mm and a blade root radius of 50 mm, the tip angle would be 28°. The root angle would be 72.5°.

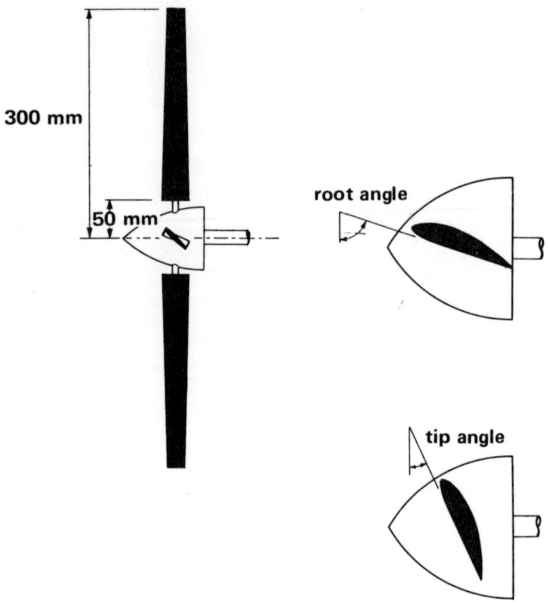

WIND DIRECTION

The wind direction could be used to excite an electromagnet which would push a brush or ball-point pen on to a drum driven by a clock. A more elaborate device would be needed for a 24 hour record.

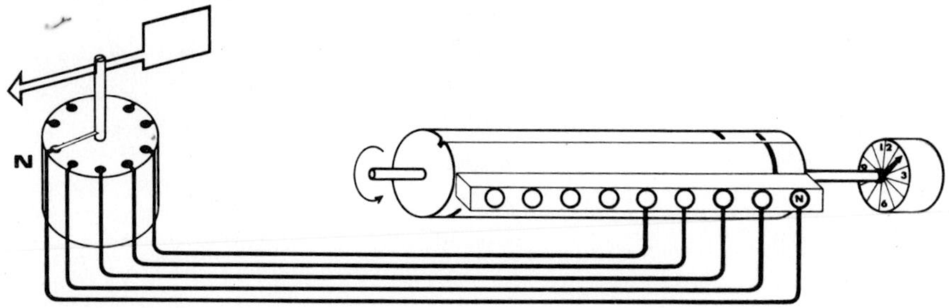